BODIES AND SPACE IN ARCHITECTURAL DRAWING

DRAWING IN

Series editors:

Russell Marshall, Marsha Meskimmon and Phil Sawdon
Loughborough University, UK

'Thinking through drawing' has become a ubiquitous trope across the arts, sciences and humanities. The rich vein of thinking, making and visualizing through drawing that is being developed across these diverse fields affords an opportunity for sustained intellectual dialogues to emerge within, between or without traditional disciplinary boundaries. The Drawing In series provides a space for new perspectives and critical approaches in the field of drawing to be brought together and explored.

In the series:

Drawing Difference, Marsha Meskimmon and Phil Sawdon
Drawing Investigations, Sarah Casey and Gerry Davies
Performance Drawing, Maryclare Foá, Jane Grisewood, Birgitta Hosea and Carali McCall
Scenographic Design Drawing, Sue Field
Serial Drawing, Joe Graham
Reportage Drawing, Louis Netter
Anatomical Drawing, Sue Field
Drawing Analogies, David Burrows, John Cussans, Dean Kenning and Mary Yacoob
Drawing Music, Marking Time, David Griffin
Drawing as Placemaking, Simon Woolham and Jill Journeaux

BODIES AND SPACE IN ARCHITECTURAL DRAWING

Line, Movement and Scale

MARIAN MACKEN AND
CARL DOUGLAS

BLOOMSBURY VISUAL ARTS
LONDON • NEW YORK • OXFORD • NEW DELHI • SYDNEY

BLOOMSBURY VISUAL ARTS
Bloomsbury Publishing Plc, 50 Bedford Square, London, WC1B 3DP, UK
Bloomsbury Publishing Inc, 1359 Broadway, New York, NY 10018, USA
Bloomsbury Publishing Ireland, 29 Earlsfort Terrace, Dublin 2, D02 AY28, Ireland

BLOOMSBURY, BLOOMSBURY VISUAL ARTS and the Diana logo are trademarks of Bloomsbury Publishing Plc

First published in Great Britain 2026

Cover image: *Tracing the collapse* (model on overhead projector), 2023, Daniel Ho. Photograph by Daniel Ho.

A catalogue record for this book is available from the British Library.

A catalog record for this book is available from the Library of Congress.

ISBN: HB: 978-1-35038-558-0
ePDF: 978-1-35038-559-7
eBook: 978-1-35038-560-3

Series: Drawing In

Typeset by Deanta Global Publishing Services, Chennai, India
Printed and bound in Great Britain

For product safety related questions contact productsafety@bloomsbury.com.

To find out more about our authors and books visit www.bloomsbury.com and sign up for our newsletters.

CONTENTS

FIGURES

ACKNOWLEDGEMENTS

We would like to thank:

Sarah Treadwell, Ross Jenner, Simon Twose and Lynda Simmons for their advice and suggestions; Leith Macfarlane for research assistance; the students, artists and designers who have kindly shared their work with us; participants at Ngā Tūtaki Encounters – Architecture Day, at the AAANZ Conference in 2019, where some of the ideas for this book were first trialled; and Waipapa Taumata Rau The University of Auckland, and Te Wānanga Aronui o Tāmaki Makau Rau Auckland University of Technology, for financial support.

Marian would also like to thank Aaron Paterson and Sarosh Mulla for the ongoing rich discussions, and María Fullaondo (1970–2023), whose passion for drawing is embedded in the work.

Carl would like to acknowledge his colleagues in the Spatial Design department at AUT for many valuable conversations and Soana for perspective and patience.

This research was carried out in Tāmaki Makaurau Auckland, and we acknowledge iwi and hapū who hold mana whenua over this rohe, especially Ngāti Whātua Ōrākei.

INTRODUCTION

ARCHITECTURAL BODIES

Drawing as a situated bodily activity

Bodies and Space in Architectural Drawing concerns architectural drawing as a situated bodily activity. It collects instances of architectural drawings for which the place and time of their production, and the physical presence of the people who make and experience them, matter, and speculates that there are more of these than is often recognized. We draw for many reasons; sometimes the body and situation of the drawer are not terribly important. But when drawing for spatial imagination and sense-making, we argue that the body provides a critical lens. How can the performative physical dimension of drawing matter architecturally? And why is it important now?

We dig into histories of architectural drawing that show how physical tools and techniques have conditioned architectural practice from the Renaissance to the late twentieth century.[1] We join these to accounts of artists' drawings in the latter half of the century.[2] In the ensuing interdisciplinary exchange between art and architecture, new ideas about drawing arise and feed into critical understandings and practices of architectural drawing and can help us understand certain features of the contemporary architectural landscape. We choose instances of embodied architectural drawing from the past as a way to chart present possibilities, particularly as they appear in the context of architectural speculation and education.

We make two potentially contentious assumptions. First, we assume that architectural drawings can be made outside the disciplinary bounds of architecture (to the extent that such boundaries can even be defined). Drawings can be architectural independent of whether they are made by architects or for the purposes of building. Second, we assume that the scope of what constitutes an architectural drawing is broad and includes three dimensions, time and a wide range of (sometimes unorthodox) media. We have arrived at these assumptions through our own interests at the margins of architectural practice, its overlaps with neighbouring fields, the influence of digital technologies and – perhaps most significantly – installation and performance art. We acknowledge these assumptions may not be applicable for all contexts, but making them enables us to recognize a vibrant strand of contemporary architectural practice. This practice (and hence our book) centres on exhibitions, installations and teaching studios rather than on building sites. We see these as venues of increasing importance for architecture and hope this book contributes to understanding ways in which they work.

From its origins in the European Renaissance, architectural drawing's foundational ideas and practices have been tightly interwoven with the project of Western modernity.[3] Perspectival drawing and descriptive geometry emerged in conjunction with presumptions about human subjectivity, the relationship between form and content, the amenability of the world to acts of measurement and abstraction, European cultural and racial power dynamics and extractive colonialism. Our two assumptions consciously cut across some fundamental premises of architecture conceived in this tradition. They recognize that architecture is broader than the production of buildings, enabling it to engage more meaningfully with cultures in which distinctions between building, landscape, event, object, structure and decoration do not appear crucial (or even make sense). Similarly, they emphasize that drawings do not always precede or determine buildings. We ourselves operate inside the inheritance of Western architectural traditions and most of the drawers we discuss are from Europe, North America, Australia and New Zealand. We are interested, however, in how our examination of embodied drawing practices might dissolve some of the edges of this frame from within and meet with those working from other directions and from different cultural and disciplinary contexts.

Drawing, from noun to verb

Our focus activates the places and times in which drawings are made and the particularities of the people who make and experience them. This may seem odd if architectural drawings are primarily understood as the production of codified instructions for building. In this mode of drawing, it is mostly unimportant where, when and by whom drawings are made (unless blame needs to be allocated or the architect's status as brand or auteur reinforced). But architects draw for reasons other than transmitting information. Drawing is how architects inquire into and make sense of situations and form design questions.[4] Drawings are the medium of architectural understanding, imagination, problem-solving, creativity and anticipation.[5] By drawing, architects tap into broader visual, technical and conceptual cultures of drawing. Deanna Petherbridge, a scholar of drawing in art practices, writes of drawing's extraordinary flexibility, nuance and range across many disciplines, 'for generating new ideas and for copying from others; for stimulating the visual imagination or documenting the phenomenal world; for studying precise detail or capturing and suggesting mood and emotion; for analyzing problems, codifying methods and mapping abstract concepts'.[6] Through drawing, ideas emerge from the mist, are puzzled out and clarified; conversely, they may subside back into the murk, dissolve or pose problems. Architect Stan Allen describes architectural drawing, and hence architectural practice, as 'transitive'.[7] Neither transparent carriers of information nor ends in themselves, drawings are a way to mediate between ideality and all of the messiness and inconsistencies of reality. Since they translate and transform as much as they transmit, their value cannot be determined purely in terms of informational efficiency.[8] Drawing uncovers and reveals; to work things out, it *draws* them out.[9] In short, drawing is a *verb* rather than a *noun*.

Architectural drawing historian Paul Emmons argues against the assumption – which he calls the 'rationalist' approach – that architectural ideas form in the mind and are only then codified and made mobile by drawing.[10] The ability to discover, invent, interrogate and elucidate through drawing, he contends, centres on its relationship to the body:

> Architectural drawing is a double act of embodiment. It includes both the drafter's physical engagement with a drawing under construction as well as the imaginal body projected into the drawing of a future building. In performing a drawing, the architect feels as well as sees a design, uniting body with understanding.[11]

From this embodied perspective, drawings are not pictures so much as construction sites into which designers enter: 'Drawing rituals consecrate place on paper by inviting the architect to imaginatively inhabit the drawing . . . [as a] place within which the architect's body imaginatively moves.'[12] We do not see rationalist and embodied approaches as being in binary opposition but, rather, simultaneous possibilities that together form an interchange or interface. Drawing's investigative, imaginative, productive and catalytic efficacy derives from its oscillation between incorporeal transparency and bodily opacity. A description such as this makes drawing an event: a situation where something *happens*, not just where it is *described*. The drawer works physically to assemble the drawing in space and time.[13]

Drawing is skilful, time-consuming work, pursued by means of an array of instruments and techniques. Until the 1990s, architectural offices were typically equipped with large drafting boards and T-squares, littered with specialist equipment, such as technical pens and erasing shields, flexible curves and pantographs, lettering templates and preprinted sheets of hatching tones. By the 2020s, this equipment has been almost completely exchanged for different apparatus: LCD panels, mice, keyboards, touchscreens, virtual reality headsets, software, servers, cameras, scanners, laser cutters and 3D printers. The digital has not dematerialized architectural drawing into a world of pure data; it has led to more complex and substantial setups than ever.[14] Contemporary building information modeling (BIM) software allows for shared digital representations of a building during design, construction and operation. In spite of the promise that drawings can be generated from this shared data, an immense amount of tweaking, adjusting, over-riding, supplementing and troubleshooting is necessary. Although the apparatus has changed, drawing remains doubly embodied (in Emmons's terms), involving both corporeal and imaginative investments.

Methodologically then, we focus on drawing as an evolving *practice* rather than a product. Allen describes practice as inherently pragmatic, improvisatory, tactical and provisional, with a 'unity inferred on the basis of its ensemble of procedures, rather than a theoretical unity conferred from without by ideology or discourse'.[15] Practices are open-ended repertoires of actions, 'numerous little operations of visualization, scaling, adjustment of instruments, options' production and selection, office presentations, historical comparisons and interpretations'.[16] This repertoire is continually flexed and adapted to context. We define drawing as architectural when made by architects, but also when created by others using architectural methods, or when they directly undertake spatial investigations or involve architectural themes. This is because our interest is in what architectural drawing *does* – how it operates in mediating between idea and reality and in catalysing architectural or spatial understandings – rather than in the extent to which it reinforces a disciplinary definition of architecture. We attend to misuses, quirks, hybrids and adaptations of architectural drawing procedures, looking for implicit, unintentional and latent effects of drawings, including meanings and associations of which the practitioner may be unaware. We notice where architectural drawing becomes uncoupled from the intention to build and escapes into neighbouring fields.

We call drawings that activate the bodily presence of the drawer or viewer in situational ways *full-scale drawings*. Scale usually establishes a proportional relationship between the space *of* the drawing (where the drawer or viewer is) and the space *within* the drawing (which it describes or refers to). It enables the fiction that another place can be apprehended through a drawing or model. By the term *full-scale drawing*, we draw attention to the way the drawing can enable existing physical situations to open up to other places. This might happen by making measured drawings at life-size – that is, metrically equivalent.[17] It can also happen, we argue, by activating the direct physical presence of human bodies in other ways so that the drawing is literally, as well as imaginatively, habitable. Instead of looking past drawings to architecture, we look to how the full-scale drawing is *already* architectural.

The remit of the *Drawing In* series is to consider the notion of thinking through drawing. What does it mean to think architecturally through drawing? And what might be gained for architectural inquiry by recognizing drawing as a situated bodily activity? It has become common to suggest that something was lost in the transformation of architectural drawing by computers from the 1990s onwards. Is this a crucial loss, a net gain or a trade-off? Has the human body been sidelined in the process? Or has it enabled a different kind of bodily involvement? We notice that computation long precedes computers in architectural drawing and tend to highlight continuities of architectural thought rather than disruptions. We hope this book points to some of the new, creative and critical ways computational drawing has been used to extend the embodied architectural imagination. A greater risk, as we see it, is in dismissing experimental drawing and interdisciplinary inquiry from architectural education. As teachers of architectural media and spatial design we find speculative practices are often pushed to the side by narrowly defined core professional competencies. In the latter part of this book, we advocate for the benefits of learning to think through full-scale drawing.

More broadly, we critically question architecture's normative assumptions about the body and support the diversification of architectural thought and practice. This has been proposed from the perspective of gendered, racialized and disabled bodies.[18] Architecture's bodies often appear abstracted and neutralized – figures for scale or ergonomic dummies. Our contribution to this diversification is in foregrounding the ways architectural drawing might directly address particular human bodies rather than generalized ciphers. Similarly, architectural drawing arises from and perpetuates Western theories of representation, visuality and space.[19] It is necessary to constantly reckon with the ideologies and assumptions embedded in our tools and techniques. It is beyond the scope of a single book to overturn these deeply embedded conceptual assumptions, but we hope to add our voices in support of those working directly to uphold alternatives.

Three themes echo through our discussion: *line*, *movement* and *scale*. While we often take an interdisciplinary approach, we work with concepts, techniques, traditions and themes that have specific architectural resonance and history. Even when we extend these terms or deviate from their traditional architectural uses, they are the paths that have guided us and by which we seek to find our way ahead. To position ourselves, then, let us reflect on these themes one at a time.

Lines and linearity

It is usual to identify lines as a defining feature of drawing.[20] The oldest senses of 'draw' concern *pulling*, suggesting that the essential feature of drawing is dragging a tool over a surface. Since the European Renaissance, architecture has been entangled with the idea that lines create and order space. In the mid-fifteenth century, Leon Battista Alberti (1404–72) opened his treatise *On the Art of Building* by defining architecture as a practice of *lineamenti*, or guiding lines.[21] These lines are at various times marks on a surface; physical contours or boundaries; and latent, immaterial or ideal relationships. He assigns the architect the task of 'finding the correct, infallible way of joining and fitting together those lines and angles which define and enclose the surfaces of the building'.[22] These lines prescribe and describe all parts of a building. Design, or *disegno*, was an intellectual labour using graphic means to define the underlying order of a building on paper prior to its materialization on the construction site.[23] According to Alberti, architecture was an ideal practice and the line was the crucial instrument mediating and fixing relationships between reality and ideality.

In the Renaissance, architecture moved from the construction site to the studio. But this was not an absolute separation from the physical labour of building. Drafting techniques developed in the Renaissance bear memories of construction activities; Marco Frascari (1945–2013) explains:

> Alberti's understanding of denoting lines (*lineamenta*) derives from the use of tracing lines, ropes, ribbons, strings or threads mostly made with flax fibres and used by the builder on construction sites during the facture of buildings. These tracing lines range from the lines pulled among pickets and the battered boards to map out the footprint of the building on the ground, to the taut strings that, stretched at the end with weights, guide the masons in their erecting of stone and brick walls, from the plumb lines directing the builders in the making of vertical walls or vertical placing of building elements, to the coloured snap lines (chalked strings used to make long, straight lines on horizontal or vertical surfaces) used to guide the layout of finishes and from measuring ropes and tapes to assess modularity, to the folding of knotted ropes to generate angles and basic proportional ratios.[24]

The building as represented on the drawing board was delineated analogically: the lines drawn across a plan pre-enact the builder's tethered and stretched string-lines. Emmons describes how sheets of paper were prepared, smoothed and laid out like sites; how squares and straight-edges were scaled-down versions of construction tools; and how the order of drawing reflected the order of operations in building.[25] Conventions, tools and procedures in common use today still hold traces of these origins. When we employ them, argues Emmons, we tap into a reservoir of disciplinary practices and knowledge founded in corporeal experience.[26] The scaled studio drawing retains its links to bodily action via full-scale acts of drawing on site.

Linearity is not only a graphic matter but also a conceptual model: an ideology or imaginary that organizes architectural practice. In spite of their apparent clarity and directness, there is nothing simple about lines in architecture. In fact, as

Catherine Ingraham argued, they have contradictory qualities, as both a kind of originary marking apparatus – since it is impossible, she says, to design anything without thinking the line first – and their use to integrate the conceptual accretions of architecture at the end of a design process.[27] On the one hand, the line is defined as 'pure extension without breadth or depth, without horizon' and on the other, simultaneously, it is the habitat of the accumulation of conceptual thinking, holding ideas that are representative of potential full-scale manifestation.[28] In *Architecture and the Burdens of Linearity*, Ingraham appraises Le Corbusier's (1887–1965) lament that European cities exemplify 'donkey urbanism' in their winding and twisting streets that follow paths originally laid down by meandering pack animals seeking the easiest route.[29] Le Corbusier believed cities could never be modern, Ingraham continues, without a rational, linear order:

> Orthogonality is Le Corbusier's theory about how to win architecture and the city away from the irrational forces arrayed within them. It extends beyond the (merely) rectilinear to any form that is erected against the monstrous, speechless, wandering, pathless incoherence of the genealogical 'line' itself – the (chaotic) history of the city and architecture.[30]

Le Corbusier imagined the straight line as a way to escape historical particularity towards an intellectual rationality capable of providing atemporal order. Anthropologist Tim Ingold further reminds us of the tensions and contradictory qualities of lines, writing of the difference between 'wayfaring' lines and lines of 'transport'.[31] A wayfarer, like a donkey, moves through space, experiencing the world continuously unfolding as they navigate and reorient based on changing conditions along their journey. A passenger, on the other hand, embarks on a train trip at one station and is whisked to their destination at another, while the landscape traversed is a blur seen through the window. The wandering line has a history. Each of its irregularities indexes some particular condition of its drawing: the wobble of a hand, the grain of a surface or a moment's hesitation. In comparison, the straight line can be specified in terms of its start and end as a direct projection across space. Straightness is not only a particular kind of line, it is the ideal of directness and rational synopsis.

In architecture, direct linearity is also associated with notions of projection: the idea that the drawing points somewhere else. Architects project spatial propositions to be constructed somewhere else and in the future. Architectural academic Ross Jenner describes architectural practice as 'something *thrown forward*, involving a form of representation awaiting existence . . . a statement of intention towards some artefact other than itself'.[32] Robin Evans takes this projective idea of architecture to be organized by the logic of geometric projection, whereby 'organized arrays of imaginary straight lines pass through the drawing to corresponding parts of the thing represented by the drawing'.[33] Drawn lines are trajectories, leaps across space and time, or hinges linking two or more places. To achieve this, lines must become replicable and transferable between drawing board and site. The linear wall as constructed, according to historian and critic Anthony Vidler (1941–2023), 'carries the freight of the line'.[34]

The drawings we consider in this book call into doubt assumptions about the linearity of architectural practice and the rational abstraction of the orthogonal line. Drawing, as the reader will encounter it here, is often decoupled from building production: it is not necessarily *towards* building. Instead, it may come *from* buildings, disclosing past lives and uses; take place *in* the built environment, yielding new understanding of existing spaces; or be reflexively *of* drawing, as a means of self-critique. Without a clear landing site as a building, it disrupts the idea of one-way projection, replacing it with networks and kaleidoscopes of potential.[35]

The logic of the line underpins computation, so we could say that computational drawing precedes the computer. By 1799, Gaspard Monge's (1746–1818) *Descriptive Geometry* had crystallized a rigorous drawing system through which any form and volume could be described, measured, calculated and manipulated. The roots of this idea of drawing as spatial calculation run back to Alberti and beyond.[36] From the 1970s, computer-mediated drawing gave these techniques new precision: digital lines were unvarying, replicable and (as Euclid proposed) without thickness or deviation. Screen-based systems effectively turned the page into an infinite plane and allowed the drafter to zoom in and out without losing resolution. From two-dimensional beginnings, computer-aided design became three-dimensional; and then, in a further step, the digital model was reconceived as a database containing performance values, costs and material qualities.[37] Now, programmable design tools and artificial intelligence are employed in the conception and development of projects. Building information management software integrates design, project management, construction processes and building performance; and computer-based visualizations allow us to experience virtual reality simulations of proposed spaces. Drawn lines are boiled down to data and supplanted by calculated ones.

It would be wrong, though, to see the rise of computation as science swallowing art. On the contrary, Matthew Allen points to the importance of avant-garde art to understanding the digitization of architecture: 'By the time the computer arrived on the architectural scene, its place had been prepared by decades of avant-gardist experimentation.'[38] In the search for an aesthetics of technology, artists had begun to conceive of artworks that acted like generative machines: László Moholy-Nagy's (1895–1946) *Light Space Modulator* (1930) was a machine generating drawings from shadows (see Figure 0.1); and algorithmically generated plotter drawings, like Vera Molnar's (1924–2023) *Interruptions* (1969) (see Figure 0.2), imagined drawing as code to be graphically executed by the computer.

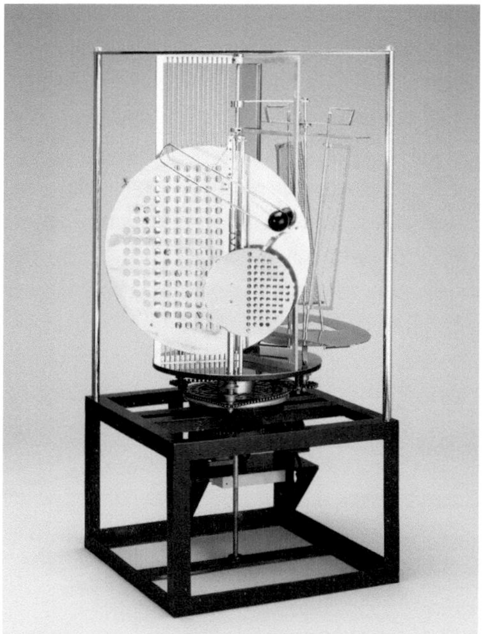

Figure 0.1 *Light Prop for an Electric Stage (Light-Space Modulator)*, 1930, László Moholy-Nagy. Harvard Art Museums/Busch-Reisinger Museum, Gift of Sibyl Moholy-Nagy. Photo: © President and Fellows of Harvard College.

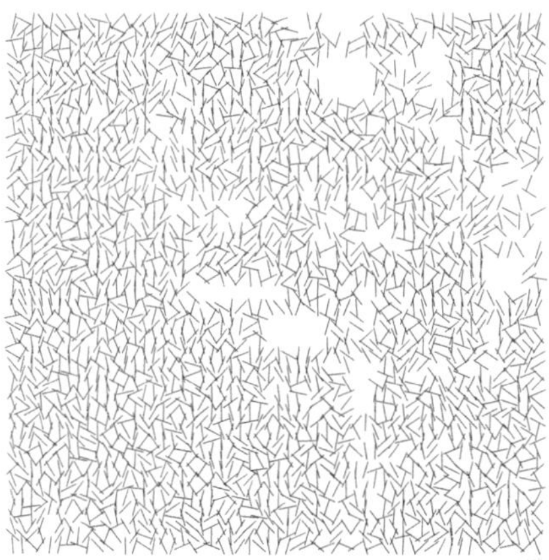

Figure 0.2 *Interruptions*, 1969, Vera Molnar. © Galerie Oniris – Florent Paumelle for the estate of Vera Molnar. Image courtesy of Victoria and Albert Museum, London.

John May argues that architecture, now 'postorthographic', creates buildings using 'simulated orthography' and, as a result, architectural drawing has become a practice of pixel-based 'imaging' rather than line drawing.[39] We elect to see this as a transformation of drawing rather than its end. The contemporary lines of architecture may be flows, code, simulations or images as much as orthographic plans or elevations. We argue these experimental cultures of practice – which now incorporate computers, digital fabrication, imaging and programming – nonetheless maintain their embodied nature and remain practices of drawing. A narrative of technological disruption can obscure important continuities and evolutionary developments in drawing.

The line, and linearity as a continuing ideology and structure of practice, recurs as a key theme in this book. Lines provide a means to move between the ideal and actual, surface and space, speculation and production, synopsis and narrative, leap and meandering walk, continuities and discontinuities. Lines divide, connect, delimit, trace and orient. They arrange ambiguous patterns, organize tensions and fracture instabilities. But, above all, they are *active*. Nancy Foote writes of this restlessness:

> Line needs space to stretch its legs. When it encounters a flat surface it is happiest running back and forth, either marking off neat sections or filling it up with random tracks. Sometimes it lies down – to rest, as it were – but it always seems to dream of marking and dividing, never of flat emptiness.[40]

Bodies in motion

Our second major theme is the moving, animate body. Architecture is typically designed to accommodate the human body, but often seems to flatten, normalize and subdue it. The body has also long served as a crucial resource for imagining aesthetics, order and place. Vitruvius, for example, has little to say about the practical accommodation of bodies, but much to say about the body as a model of proportion and order.[41] Embedded in this paradigm is the idea of the body as a model for the relationship between parts and wholes. As the body was understood to be organized into functional elements (hands, eyes, organs) it was also understood to be brought together by the animating principle of the intelligence. In Vitruvian architectural theory, and in Roman thought more generally, wholes were *tripartite*: the column was understood to be complete and to stand like an idealized body because it consisted of a base, shaft and capital.[42]

In the Renaissance, Alberti and other architectural writers carried forward the idea of the body as an ordering principle.[43] Bodies were generalized and abstracted as *the body*, with a definitive, singular sense that implies universality and takes away from the consideration of particular bodies. Proportional systems in Renaissance architecture, to which the nascent standards of architectural drawings were oriented, sought harmonious resonance with universal and eternal cosmic order.[44]

The displacement of particular bodies in favour of general ones persists in the impersonality of most contemporary architectural drawings. Diana Agrest points out that the codes, conventions and graphic styles of professional practice actively suppress the particularities of the drafter in their work: 'The body, which stands metonymically between the world and representation through the eye, is either repressed through numerical relationships or reduced to an eye/mind mechanism.'[45] Qualities specific to the drawer – such as variations in the thickness of a hand-drawn line that might indicate emphasis or tentativeness – are eliminated, regulated by drawing instruments, software and templates. The drawer's presence may only be hinted at by their initials in the title block or by inadvertent smudges and smears.[46] In addition, working drawings come from many hands – designers, assistants, drafters, consultants – and so, according to Christine Macy, are contemporary palimpsests.[47] They are imagined to be faithful transcriptions of architectural intentions and strict guides for the labouring bodies who will carry out these intentions. In philosopher Nelson Goodman's terms, architectural drawing is 'allographic': a descriptive notation that does not derive its creative value from being an authenticated original.[48] The architectural sketch, meanwhile, has been elevated to a kind of autographic signature, a distillation of the individual architect's insight and sign of authenticity.[49]

Sketches also activate bodily experience: it is possible to see how a particular curve or spacing of elements *feels* by drawing it.[50] At the same time, architects project miniature versions of themselves into their drawings, imaginatively walking through plans, foreseeing bodily activities of future inhabitants, projecting how spaces might be experienced even while the drawing is incomplete. The doubled body of the drafter is simultaneously physically present in the experiential act of drawing and projected imaginatively into the drawing to anticipate its construction and inhabitation.

Architecturally projected bodies tend to be individual, normative, static and generalized. Anthropometrics and ergonomics define bodies statistically, offering typical dimensions and ranges as design data.[51] This extends to studies of organic performance: how bodies respond to environmental conditions such as temperature, humidity and light. The dimensions, sensitivities and capabilities of bodies are abstracted in the form of normalized statistical models. Newer theoretical constructs, such as accessibility and universal design, acknowledge the limitations of this approach and the fundamental diversity of human bodies. In the field of disability studies, the variable forms and capacities of bodies are recognized as embedded in social, political and ethical contexts.[52] This requires a rethinking of how those capacities are *defined* or even *invented* by their contexts rather than merely being made visible by them. Critically, these studies shift away from tendencies to normalize conditions towards demonstrating how these capabilities can be attended to by architecture.

Bodies, of course, are not just things we consider from outside; we are ourselves bodies and know them from the inside out. Immanuel Kant (1724–1804) argued that our abstract understanding of three-dimensional space could only come from our concrete experience of embodiment. We stand vertically, and the asymmetries of our body – left-right, up-down, front-back – generate our concepts of space as being three-dimensional.[53] As Edward S. Casey put it, 'The body is the pivot around which the three dimensions of spatial extension arrange themselves. . . . Without the body's lopsidedly two-sided ingressions into particular regions and places, space would be merely a neutral, absolute block or else a tangled skein of pure relations built up from pure positions.'[54] Twentieth-century phenomenology built on this understanding. Maurice Merleau-Ponty (1908–61) pointed out that 'my whole body for me is not an assemblage of organs juxtaposed in space. I am in undivided possession of it.'[55] We are able to apprehend other things to the extent that 'every figure stands out against the double horizon of external and bodily space'.[56] For Merleau-Ponty, geometry does not underpin the body; it is only revealed in bodily terms. Isometric and isotropic space is an imaginary abstraction. They only appear fixed and unchanging in relation to our own unceasing bodily movement and change.

To inhabit space then is not just to be positioned or contained. Inhabitation implies action. Phenomenology is frequently misunderstood (particularly in architectural circles) as predominantly to do with the qualities of sensory experience; for Merleau-Ponty, it is deeply entangled with intention, use and imagination: 'My body is wherever it has something to do.'[57] The body is a context for habit, familiarity and memory. Because of our prior experiences and imagination, we are able to find ourselves in places laden with possibilities for action. It is not only architects who inhabit space in imagination; we are all doing it constantly. The body is, according to Casey, 'a source of intentionality and projects, correlations and orientations'.[58] Space forms around us in relation not only to the form and organization of our bodies but also to our activities and our continually alert sense of what it is possible for us to *do*.

Undermining the Western philosophical and cultural split between mind and body, Merleau-Ponty came into dialogue with non-Western understandings. Japanese philosopher Kitarō Nishida (1870–1945) described the self as something that *resolves* out of an ongoing bodily encounter. Experience, an 'intimate corporeal resonance' is prior to the sense that one's body is separate from the things one perceives and interacts with.[59] The body, according to Nishida, is '*shintai* . . . the immaculate iconic tool of nothingness in its full commitment to accept the world and to transform the world by creating new figures'.[60] Doing and transforming might also involve negating and stilling.

Even where it has engaged non-Western perspectives, phenomenology may present a too-tidy account of the body. Contemporary feminist writers and theorists of art have emphasized instead the body's disjunctive, disruptive and subversive potential. For Elizabeth Grosz, the body is at once entirely concrete and absolutely open-ended:

> By 'body' I understand a concrete, material, animate organization of flesh, organs, nerves, and skeletal structure, which are given a unity, cohesiveness, and form through the psychical and social inscription of the body's surface. The body is, so to speak, organically, biologically 'incomplete'; it is indeterminate, amorphous, a series of uncoordinated potentialities.[61]

Bodies are complexes of subjectivity and capacity, 'centers of perspective, insight, reflection, desire, agency'.[62] They are also cultural, social and discursive with all sorts of contextual entanglements. Perhaps the body is a kind of hinge, a pivot around which things turn, an articulation of the biological, psychic and social. It may open towards these contexts or turn inwards away from them. Dramatist José Gil writes of a body that 'opens and shuts, that endlessly connects with other bodies and elements . . . traversed by the most exuberant fluxes of life . . . That is: a *paradoxical body*.'[63] Dancers and performers, he argues, do not move in space; rather, they secrete or create space with their movement. In performance, '[t]he body is at once the most solid, the most elusive, illusory, concrete, metaphorical, ever present and ever distant thing – a site, an instrument, an environment, a singularity and a multiplicity'.[64]

These complex and contingent descriptions blur lines that architecture has traditionally maintained between site and inhabitant, mind and body, meaning and use. Even if it were possible to synthesize a single theory of the body, we certainly do not attempt one here. Rather, for us, body (like line) is an architectural *question*. Just as we redirect from line to questions about linearity, we also shift from body to questions about corporeality and the ways human bodies enact, inhabit and activate drawings.

Scale and situation

Our third theme is *scale*, which we similarly reorient as questions about *scaling*. In the rational world of geometric drawing, scaling is a proportional transformation of size: 'a mathematical function applied as a technical procedure'.[65] Two- and three-dimensional shapes can be made proportionally larger or smaller by a given ratio. In architecture, scale is crucial for imagining and describing things much too large to fit onto a sheet of paper. An object can be translated into the world of the architectural drawing by being measured and made wieldy by a proportional reduction in size. Certain fixed ratios have become associated with particular kinds of drawing and levels of detail. So, a 1:500 site plan will show arrangements of buildings in relation to the local area, a 1:100 scale plan will show a composition of spaces and furniture and a 1:10 detail reveals how windows are fitted. Larger scales encourage an increase in detail, whereas smaller ones encourage simplification. Common scales are conventions that facilitate comparison. Where imperial measurements are usual, different scales are also common: for example, 1:12 representing one inch to a foot. Where converting between scales is not important – with toys for examples – other unusual scales are used. The most common model railroad scale (HO) is 1:87, military models are often at 1:72 and Hot Wheels® cars are approximately 1:63. Scale separates these worlds from one another. Over time the users of scaled drawings become familiar with certain scales and acquire an intuitive sense of size.

Scale is a bodily relationship that brings things within the scope of a designer's attention and action. The physical size of the drawing in relation to the drawer matters. Enric Miralles (1955–2000) insisted that conceptual drawings needed to be small, made on pieces of tracing paper cut to size, so that the relative size of the drawing in relation to the gestures of the hand facilitated diagrammatic clarity.[66] Scale relates to the scope of human bodies: the reach of their arms, the resolution of their vision, their capacity to carry. The term *scale* recalls the Latin *scāla*, suggesting 'a stair providing the means for ascending and descending between the great and the small'.[67] Scale, architecturally used, is also a means to interrogate a design. As Albena Yaneva sees it, scaling is an 'experimental situation' that 'relies on procedures for partial seeing: scoping, rescaling, extending and reducing'.[68]

Digital models handle scale fluidly. The model is specified at full size in the virtual environment and only assigned a scale once printed in the physical world. Other mechanisms such as layers, versions or resolution control permit filtering levels of detail. In digital drawing and imaging, scale is instead reconfigured as a zoom. Rather than stepping up and down the scalar stair, the zoom imagines that all detail exists at once, but only becomes visible through relative proximity or invisible with distance. In this logic, the large-scaled view is a view from afar. Ray and Charles Eames's (1912–88; 1907–78) film *Powers of Ten* (1977) prefigured this paradigm, which has become normalized as a prevailing visual assumption.[69] Through using online mapping services, we have become accustomed to aerial photographs stitched together as if they were a single scaled visual field; the apparent smoothness of the zoom masks discontinuities and variations in the data.

Architectural scale figures endow drawings with a sense of relative size, inviting the viewer to insert themselves into the drawing's world. They are often simplified and non-specific: bodies perform graphic functions, interchangeable with the viewer's imaginary body. In *An Unfinished Encyclopedia of Scale Figures Without Architecture*, Michael Meredith and Hilary Sample extracted 1,000 instances of scale figures from architectural drawings by twenty different architects.[70] Isolated from their architectural context and arranged alphabetically by the architect's name, they become signatures expressing the architect's style rather than the physical identity of occupants. They routinely lack distinguishing features of character (faceless or turned away from the viewer), and are sometimes highly abstract, resembling pegs, elongated diamonds or blobs. Perhaps faces would divert attention or imply the existence of a particular subjective experience; it is easier for the viewer to imaginatively slip into a more generic figure.

Digital drawings and rendered images may include generalized scale figures – perhaps ghosted or outlined – populated using collections of preprepared photographic figures or posable three-dimensional bodies. These figures migrate from project to project. Teachers of architecture will often recognize certain figures turning up again in student designs like familiar acquaintances. Michael Young argues modern scale figures originate in eighteenth-century picturesque aesthetics: 'The human figure was indeed an essential element of the picturesque image but did not serve as the narrative subject, as in neoclassical painting. People in a picturesque painting are instead objects: they give scale, lend familiarity, and aid in visually domesticating the landscape.'[71] Such bodies, like plants, weather, furniture and other objects, make up the drawing's 'entourage': things not designed by the architect but intended to guide the reading of the image.[72] These can imply the presence of human life; beds, dining tables, pot-plants or pets normalize spaces: we sleep here, we eat in this way. Independently of whether the image has a metric scale, entourage places it within the reach of our bodily imagination, rendering it a relatable scene for action.

Anthropologists have used the term *scaling* to describe the various ways we imagine and negotiate our relationship to things far outside the scope of our bodies. According to climate historian Deborah Coen, scaling is

> [H]ow we think, for instance, about how one individual's vote might influence a national election, or whether buying a hybrid car might slow global warming. It can also be a way of situating the known world in relation to times or places that are distant or otherwise inaccessible to direct experience.[73]

Scaling in this sense is not a neutral practice. Anthropologist Anna Tsing describes how it was fundamental to colonialism and capitalism. These expansionist enterprises 'develop the quality called scalability, that is, the ability to expand – and expand, and expand – without rethinking basic elements'.[74] Standardized logistics, fungible and disposable workers, exchange markets and the idea of land as interchangeable *terra nullius* all rely on a scalar imagination. It may seem overzealous to connect the apparently benign and straightforward practice of drawing to scale with such immense socio-political constructs. Scale, though, asks exactly this kind of question: how do we understand and imagine the relationship between the very large (or the very small) and the space of our own bodies and actions?

As a way of situating, scale is closely allied to the architectural idea of *site*. Sites are not only 'spatially finite places' but also involve the selection, interpretation and positing of contexts.[75] Each project involves a determination about what attributes are pertinent. In writing of Le Corbusier, Mark Wigley proposes that 'the project is never more than the construction of a site'.[76] Although notionally site precedes project, it is also defined in terms that would make it amenable to the project. In Le Corbusier's case, rapidly sketched diagrams articulated sites in terms of formal and compositional tensions: horizon versus a rumpled line of hills, or the window versus the open terrain. Architecture mediated between distant and proximate, inside and outside. He explicitly imagined site as a matter of extending one's (colonial) grasp: 'Your plans heretofore confined to the few square feet of a room, can now be extended until they form an empire stretching to the outer limits of horizons. Revealed to you, you can conquer them.'[77] Site and scale involve a relationship to horizons of vision, action, comprehension and imagination.

Site has proved a potent term in art practice, attested by the rise, since the 1960s, of terms like '[s]ite-determined, site-oriented, site-referenced, site-conscious, site-responsive, site-related'.[78] These interlock with other scalar themes like environment, context, setting, the urban and landscape. Art historian Miwon Kwon describes how, throughout the twentieth century, site – initially understood as an 'agglomeration of the actual physical attributes of a particular location' – comes to signify 'a relay or network of interrelated spaces and economies', and eventually 'broader cultural, social, and discursive fields . . . organised intertextually through the nomadic movement of the artist'.[79] Each of these sites involves relations to larger entities, outside the direct scope of the work, but necessary to it.

When we discuss scale in this book, then, we have in mind more than simply the conventional practice of proportionally reducing the dimensions of drawings or models. Scale, for us, concerns inhabitation, occupation, use and meaning – the sense given to space in relation to human bodies and activities. The sense of scale is something narrated, felt or enacted, not simply measured. The sites of architectural drawing, as we have already noted, are often doubled (or multiply layered). There is the here-and-now of the drawing's production and the bodies of drawer and viewer, and an elsewhere-and-elsewhen to which it points. Many of the works we will examine complicate this division, and nuanced manipulations and questions of scale arise as they situate themselves.

Looking down the line

Line and linearity, body and movement, scale and situation – these themes tangle together in various ways in the chapters ahead. Our aim is not to give them fixed meanings, but to indicate how they work in practice: the questions they raise, the way they organize or disrupt architectural drawing and how they resonate in contemporary architectural practice. We are also interested in how they operate at the transformative margins of the discipline. Although our inquiry is centred on architecture, many of our examples are drawn from related spatial fields: art, dance, performance and other areas of design. Our hope is that, by exploring edge-cases and unconventional examples, we may cast new light on existing architectural practices and indicate possibilities for new ways of practising.

In our opening chapter, we focus on a moment in architectural history when expanded ideas about drawing prompted new questions and problems for architecture. In the 1970s, interest in architectural speculation intensified and was allied to the rising importance of exhibitions as a mode of architectural practice.[80] In this period, exhibitions began to be seen as themselves architectural, rather than simply representations of architecture. At a similar time, drawing as an art practice changed through its dialogue with two new genres of art: installation and performance art. Drawing, and especially the idea of drawing as a material enactment (rather than a means of picturing), was a significant element of these new genres. Curators Cornelia H. Butler and Catherine de Zegher describe how drawing 'emerged from the first half of the twentieth century as a field coextensive with real space, no longer outlining an illusion marked off from the world'.[81] From this perspective, we compare three architectural exhibitions in New York – *Idea as Model* (1976), *Architecture I* (1977) and *Architecture II* (1980) – with the contemporaneous and influential art exhibition *Line as Language* (1974), curated by Rosalind Krauss. We point to an exchange of ideas about the nature of drawing and body leading to a new emphasis on architectural drawing as a physically embodied practice during the 1990s and early 2000s.

Chapter 2 concerns drawing technologies and the idea of the full-scale drawing. In the 1990s, a relatively settled (but by no means homogeneous) body of graphic techniques was disrupted and reconfigured by paradigms of digital simulation and programming. But computation, in the form of the practical techniques of descriptive geometry, had long been central to architectural drawing. Using this technical system, views of three-dimensional objects could be calculated and virtual objects measured, studied and transformed. This chapter tests our understanding of architectural drawing as an embodied spatial practice by reflecting on three drawing techniques that anticipated computational drawing: constructing perspective, measuring human bodies and surveying terrain. Building particularly on Emmons's account of embodiment in architectural drawing, we argue that each of these techniques can be understood as the production of full-scale drawings that precede later scaled and surface-bound ones.[82] We trace these forward into contemporary digital methods of scanning and surveying, arguing that full-scale drawing is (and has long been) a means to translate between material and geometric worlds.

Our third chapter concerns drawing in time. Performative drawings by artists in the 1970s (including Jim Allen (1922–2023), Carolee Schneemann (1939–2019), Dennis Oppenheim (1938–2011) and Trisha Brown (1936–2017)) posed drawing as a temporal occurrence that can be captured in documentation or on film, but is itself inherently bound to the particular time and place of its enactment. Bodily gesture, and gestural drawings such as the sketch, have long been understood to express embodied skill, and often also assumed to capture subconscious, latent, social and emotional aspects of the drawer.[83] Early motion-capture technology – such as Lillian and Frank Gilbreth's (1878–1972; 1868–1924) time-and-motion studies of workers – recorded bodily gesture as lines in space.[84] These performative drawings were taken to reveal degrees of expertise or opportunities for efficiency. Methods of sampling and scanning that resemble techniques discussed in Chapter 2, augmented by the camera, turned into techniques for studying (and regulating) bodily movement. Architects, fascinated by incorporating time and movement into design, explored ways to capture gesture: by shaping space around bodies in three dimensions, emphasizing the freehand line, inventing choreographic notations and attending to the physical labour of construction. By the time digital drawing arrived with its moving flows, fields and swarms, the gestural line was well-established as an architectural theme. In the second part of the chapter, we make readings of a series of architectural drawings from the 2010s that work with motion capture—Sarosh Mulla, Aaron Paterson and Marian Macken's *Drawing the Room | Drawing within the Room* (2019), Níall McLaughlin and Yeoryia Manolopoulou's *Losing Myself* (2016) and *The Dream Life of Driverless Cars* by Matthew Shaw and William Trossell—noticing how they complicate the when, where and who of architectural drawing.

From performance, we turn to installation in Chapter 4. If exhibitions are shows, a temporary framing of a work or works, then installations are immersive scenes.[85] They are not so much simulations of architecture as open-ended fields of encounter activated by the physical presence of the viewer or participant. We examine six architectural works from the period 2008–20: Mark Dorrian and Adrian Hawker's *On the Surface* (2014); Sue Gallagher's *Line-dwelling* series (2016–18); *Canyon* (2018), by Simon Twose, Jules Moloney, Anastasia Globa and Lawrence Harvey; *Everest Death Zone* by James Craig and Matthew Ozga-Lawn (2013); *AirXY* (2008) by Erik Adigard and Mark Salter; and *Drawing Room* (2020) by Mulla, Paterson and Macken. Each explicitly engages with both the language of installation and an expanded understanding of architectural drawing, activating the body of the viewer as an integrating and potentially disruptive presence. We present them as exemplifying an important vein of contemporary architectural drawing practice.

In our final chapter, we draw out some of the implications of this contemporary architectural installation-drawing practice for architectural education, using examples from our own teaching experience. We take the view that student work is not only preparation for professional practice but also an important dimension of architectural inquiry. We argue that situated bodily drawing practices are particularly valuable for fostering architectural imagination, making sense of complex situations, supporting the acquisition of disciplinary knowledge and enabling disciplinary self-critique. Making and acting at full scale need not be at odds with speculation and experimentation. The studio is crucially a physical and shared space, a 'suspension of the world so it can be studied'.[86] These student examples demonstrate that practices of bodily architectural drawing are of concern to the next generation of practitioners.

Architectural drawing as an experimental, speculative and critical practice is a vital aspect of contemporary architecture. There are, of course, many important practitioners and works beyond those we discuss here; our aim is not to provide a summary of contemporary architectural drawing in general.[87] Our intention is narrower: to gather instances of architectural drawing that specifically engage with the bodily presence of drawer and viewer or participant; to chart something of their interdisciplinary genealogy; and to advocate for their critical, speculative and pedagogical value.

1
DRAWING IN SPACE
EXPANSIONS OF ARCHITECTURAL DRAWING

Architecture in reverse

In 1988, a group of students at Cranbrook Academy of Art in Michigan purchased, for one dollar, a house in Detroit slated for demolition. This was at a time when the city was in decline: the relocation of the industrial economy had led to the abandonment of neighbourhoods and demolition of buildings. Two years earlier, in 1986, no building permits had been issued in the city at all.[1] Dan Hoffman, then head of Cranbrook's Department of Architecture, described the city as being in the process of 'unbuilding'.[2] Within this context, the students undertook an analogous 'act of unmaking'[3] on the property they had bought. Over a two-week period, they slowly dismantled the house, cleaned and sorted its materials, separated and collated its constituent parts and moved these to a gallery space. Steel drums held loose materials next to neat piles of similarly dimensioned timber battens arranged amongst the gallery's columns, available to be reclaimed by the community[4] (see Figures 1.1 and1.2).

Hoffman and the Cranbrook students – James Cathcart, Frank Fantauzzi, Terence Van Elslander, Jean-Claude Azar and Michael Williams – understood their action as a reversal of linear architectural practice by moving from building to drawing: 'The project even reversed the accepted architectural practice of beginning with a two-dimensional representation, such as a plan, and expanding it into the volume of a building. By starting with a building rather than a representation, the architects discovered the plan through a process of excavation.'[5] This project, titled *9119 St. Cyril* for its location, inverted the typical architectural sequence of buildings emerging from drawings, to be a kind of

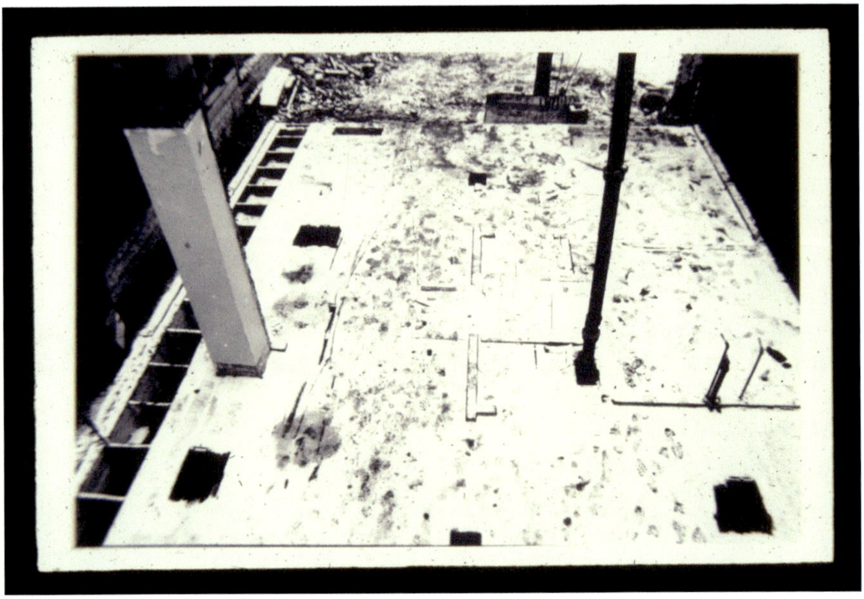

Figure 1.1 *9119 St. Cyril*, Detroit, USA, 1988, James Cathcart, Frank Fantauzzi, Terence Van Elslander, Jean-Claude Azar and Michael Williams. Photograph by Frank Fantauzzi.

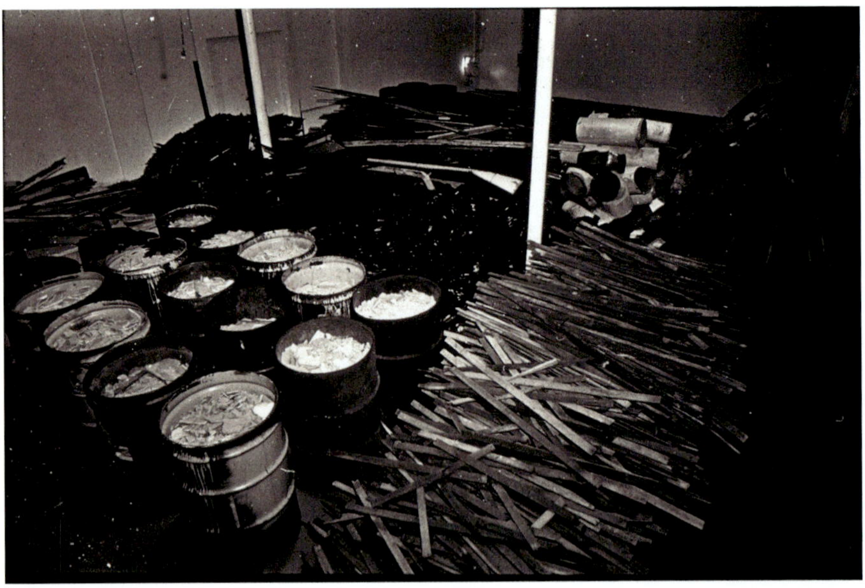

Figure 1.2 *9119 St. Cyril*, Detroit, USA, installation view, 1988, James Cathcart, Frank Fantauzzi, Terence Van Elslander, Jean-Claude Azar and Michael Williams. Photograph by Frank Fantauzzi.

'design-build in reverse'.[6] For a moment during its transformational *undoing* – collapsing volume and elevation – the ground plan was exposed. This recalls the Renaissance understanding of the plan as a physical footprint, a shape traced on the surface of the ground rather than an abstract cut.[7] Independent of the building's pre-construction scaled plan, this revealed plan resulted from the removal and then sorting and arranging of that which was built upon it and bore the marks of the building's life and use. Instead of a diagrammatic view projecting a future building, this plan was a discovery of something latent in the material world.

9119 St. Cyril aligns more easily with performative, site-specific artworks, like Robert Smithson's *Partially Buried Woodshed* (1970), Mierle Laderman Ukeles's *Hartford Wash* (1973) and Gordon Matta-Clark's (1943–78) *Conical Intersect* (1975), than with traditional ways of doing architecture. These works, with their acts of burying, washing and removing, address buildings as already-existing things on which artists labour. To the extent that they worked against the idea of architectural completion, they might be understood as anti-architectural. It may be unorthodox to call *9119 St. Cyril* a drawing in any conventional sense, but this chapter entertains such a reading. To do this, we explore the relationship between full-scale architectural experimentation and the critique of drawing as representation. We acknowledge that drawing has been a particularly fertile zone of cross-pollination between art and architecture in the second half of the twentieth century, particularly through the recognition of drawings themselves as material and spatial. By the 1990s, many instances can be found of what we refer to in this book as *full-scale drawings*: by practising architects and designers in related fields, in architecture schools and by artists. They are architectural inquiries that activate bodily presence in the form of installations, performances, or spatial constructions. These inquiries anticipate the rising importance of exhibitions, installations, performances and creative-practice-led research for architectural practice in the twenty-first century.

We centre this chapter's discussion on a series of gallery exhibitions in the United States between 1974 and 1981, noticing drawing themes, questions and practices that flow through into the 1990s (at which point they linked up with those arising from architecture's 'digital turn').[8] First, we recount drawing practices by artists, interwoven with the emergent genre of installation art. Second, we note how understandings of the architectural exhibition change in this period: rather than representing architecture that happens elsewhere, the drawings and models on display (and the show itself) were understood as autonomous artefacts. A critique of architectural representation arises from the conjunction of these two strands, centring on the productive discontinuities and paradoxes implicit in drawing. To what extent does architectural invention depend on ambiguity and paradox? Can architectural drawing be meaningfully decoupled from its role in producing future buildings? These questions, we suggest, turn on the concrete, material presence of the drawing, drawings which are situated spatially and temporally, and scaled by the bodies of drawer and viewer.

Situated drawing

In the 1960s, artistic enthusiasm for drawing flourished:

> With characteristic artistic obsession, they drew on plaster, on plastic, on printed grids, and on the earth. They drew with the body, with the typewriter, and with light; they rubbed, sprayed, blew, and pounced. They drew not only with traditional materials of graphite and charcoal, watercolour and ink, but with soot, sandpaper, gunpowder, and beeswax. The subject matter became the hidden, the mundane, or the mathematically structured field. Artists depicted images from the news, comics, and advertisements; they scavenged diagrams, instructions, lists of words or letters, and other seemingly ordinary fragments of daily life.[9]

A 1974 exhibition at Princeton University Art Museum – *Line as Language: Six Artists Draw*, curated by Rosalind Krauss – captures the new aspirations that arose from this period. Krauss affirmed drawing as a medium independent of 'illusionistic projection' or 'pointing to something beyond itself'.[10] The show featured work by six artists – Mel Bochner (1940–2025), Sol LeWitt (1928–2007), Robert Morris (1931–2018), Dorothea Rockburne, Richard Tuttle and Richard Serra (1938–2024) – seeking to evade drawing's referential tendencies by emphasizing its physical enactment.

Rockburne posed the working question for herself, 'How could drawing be of itself and not about something else?'[11] One of her contributions, *Neighbourhood* (1973), was a sheet of paper folded twice and hung on the wall (see Figure 1.3). Red, black and graphite lines marked from the sheet to the wall record the act of folding and unfolding, described by Krauss as 'residual indicators, on a flat plane, of a set of three-dimensional operations specific to and generated by the topological space of the paper'.[12] The paper was not the surface on which a drawing was made but itself a drawing instrument. *Neighbourhood* also indexes the scope, scale and activity of Rockburne's body in relation to the geometry of the paper: we comprehend her acts of reaching, holding and tracing. In her related series, *Drawing Which Makes Itself* (1973), double-faced carbon paper was used in a similar manner, so that the paper itself 'both forms itself and is formed by the act of forming. . . . The paper both receives the form and forms the work'[13] (see Figure 1.4). Fingerprints and smudges made by the drawer remained visible. These works implicated the space of the gallery: art historian

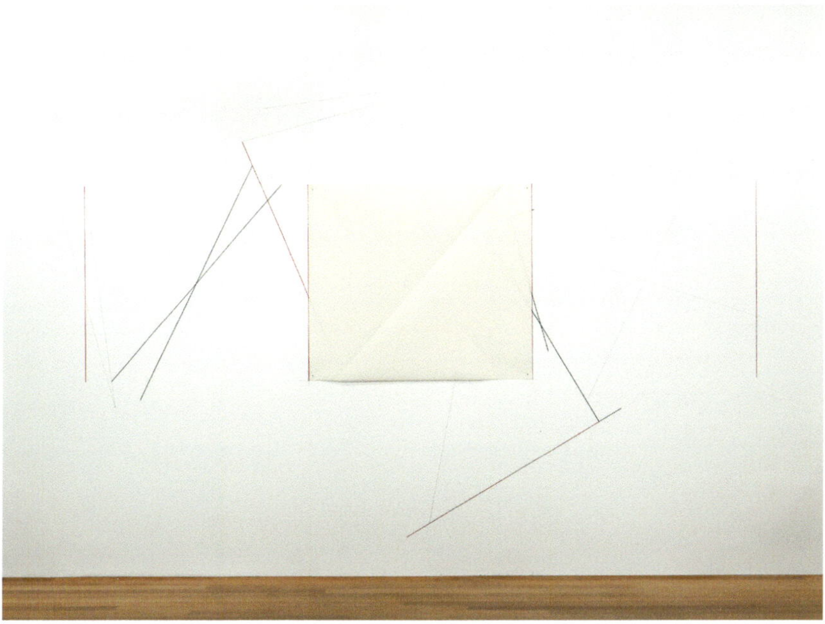

Figure 1.3 *Drawing Which Makes Itself: Neighbourhood*, 1973, Dorothea Rockburne. Transparentized paper, pencil, coloured pencil and felt-tip pen on wall, 107 × 150 in. (271.8 × 381 cm). Gift of J. Frederic Byers III. Acc. n.: 296.1978. Digital image, The Museum of Modern Art, New York/Scala, Florence. © Dorothea Rockburne. ARS/ Copyright Agency, 2024.

Anna Lovatt writes that they 'describe the floor, the walls, and the ceiling not as being floors and walls and ceiling but almost as a flat continuous surface that happened to have three dimensions'.[14]

LeWitt's contribution to the exhibition was a drawing made directly on the wall, eliminating any intermediary between the work and its support.[15] *The Location of Points: 2 Wall Drawings* was first defined as a written procedure, then executed in the gallery. The first step was straightforwardly specified as marking the exact centre of the wall's surface with a cross; the second involved a comically complex process of bisecting and connecting points to identify a location for the second cross. The two resulting crosses on the wall belie the difficulty of the drawing's production. If Rockburne's drawing was indexed to the scale of her body through acts of folding, LeWitt's was indexed to the scale of the building. LeWitt's *Wall Drawing* series (1969–2007) separates the acts of planning and execution in a way architects would find familiar.[16] Lovatt writes that one of LeWitt's *Wall Drawings* 'is structured by the architecture but also serves to articulate it, rendering the texture, size, shape and fixtures of the wall visible to

Figure 1.4 *Drawing Which Makes Itself: FPI 16*, 1973, Dorothea Rockburne. Folded paper and ink, 30 × 40 in. (76.2 × 101.5 cm). Blanchette Hooker Rockefeller Fund. Acc. n.: 303.1974. Digital image, The Museum of Modern Art, New York/Scala, Florence. © Dorothea Rockburne. ARS/Copyright Agency, 2024.

the spectator'[17] (see Figure 1.5). Elsewhere in *Line as Language*, the line peeled away from the surface and reached through space. Richard Tuttle began his *Wire Work #44* (1972, re-executed in 1974 for *Line as Language*) with a freehand pencil line on the wall. Wire, attached to small nails positioned at the line's two end points, looped into space joining them and producing a third line of shadow (see Figure 1.6).

Not included in *Line as Language*, but concurrently working in a similar vein, Fred Sandback (1943–2003) drew in space with coloured elastic rayon cord. In 1970 he exhibited two works at the Dwan Gallery in New York (see Figure 1.7). In one room, three taut white elastic cords, secured by anchor points, formed a trapezoid extending a small way out from the wall; in another, a simple blue cord spanned the room diagonally. He used the term *pedestrian space* to direct attention to the way they intervened in the viewers' movement through the gallery.[18] Art historian Reinhard Spieler writes that these taut strings seemed like 'delicate drawings that have left paper and surface to reach out across the room'[19] (see Figure 1.8). The lines react to the architecture, its light and openings, plan and volume, becoming part of the buildings that house them. After each exhibition, the threads were discarded; they became special only when held taut

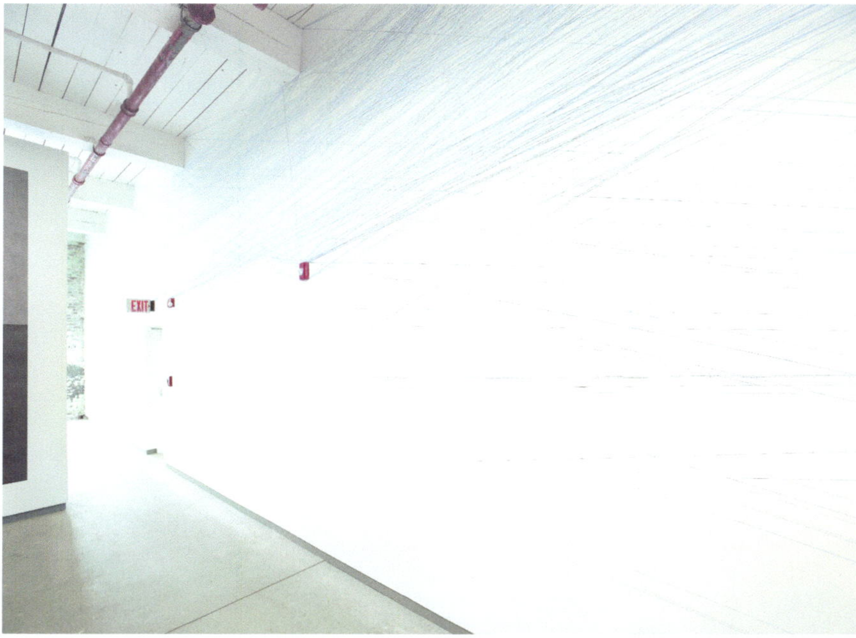

Figure 1.5 *Wall Drawing #51*, 1970, Sol LeWitt; recreated at MASS MoCA for *Sol LeWitt: A Wall Drawing Retrospective*, 2008. Photograph by Kevin Kennefick, courtesy of MASS MoCA, North Adams, MA. © Sol Lewitt. ARS/Copyright Agency, 2024.

between anchor points, doing the job of spanning between points in space.[20] Once slackened, they lost their sculptural function (see Figures 1.9 and 1.10).

Sandback and the artists of *Line as Language* express a new understanding of drawing as three-dimensional rather than surface bound. Krauss writes, in her essay accompanying the exhibition, that the six artists make 'an attack on the traditional transparency of the line – the way that line had always seemed to make available something that lay behind or beyond it'.[21] Being made manifest, the lines' meaning was affected. The lines of *Line as Language* directed attention to their material nature – wire, graphite, crease or shadow – and their temporal enactment: towards the here and now of their production, not the projected elsewhere of representation. As LeWitt put it, '[a] line only stands for itself. It doesn't describe anything but a line.'[22] In a later essay, Krauss argued that three-dimensional drawings, by addressing the physical mobility of the viewer, interrupted the distinction between figure and ground and, thus, 'figurative completion'.[23] Drawing in space was not simply the addition of a new dimension to an existing practice but a redefinition.

In the retrospective exhibition *On Line: Drawing Through the Twentieth Century* (MoMA, 2010–11), curators Cornelia H. Butler and Catherine de Zegher

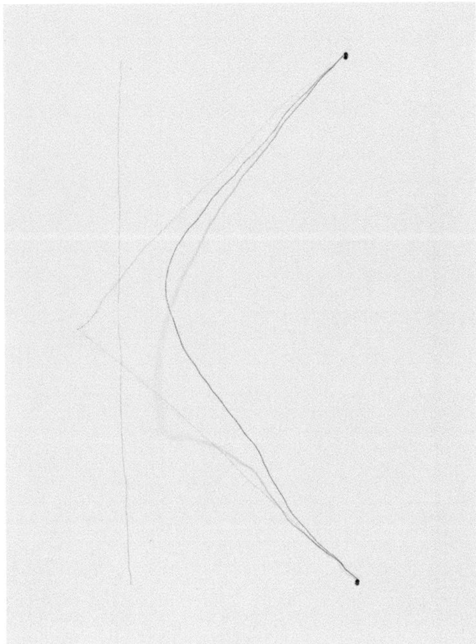

Figure 1.6 *44th Wire Piece*, 1972, Richard Tuttle. Florist wire, nails and graphite. Overall: (installed) 38 1/8 × 19 1/4 × 9 1/4 in. (119.4 × 55.9 × 28.6 cm). Other (wire): approx. 58 in. long. Acc. n.: 97.109. The Museum of Contemporary Art, Los Angeles. Gift of Lannan Foundation. Photograph by Brian Forrest.

chart this expansion and translation of drawing. De Zegher writes that as a result, drawing was redefined 'as a field coextensive with real space, no longer outlining an illusion marked off from the world'.[24] Rather than a specific medium, they argue, drawing had become a transdisciplinary concern centred on the nature of lines. Two major genres of art which emerged in the 1960s affected these investigations into the line: installation art and performance art. The term *installation* (appearing in the *Oxford English Dictionary* in 1969) describes a mode of art practice 'into which the viewer physically enters, and which is often described as "theatrical", "immersive" or "experiential"';[25] before this, the term referred to the arrangement of an exhibition (more on this in Chapter 4). Performance art (addressed in Chapter 3) shifted the nature of artistic output to event and process rather than artefact. Butler and de Zegher show that the expansion of drawing formed exchanges with both genres. El Lissitzky's (1890–1941) *Proun Room* (1923, reconstructed several times, including for *On Line*) – essentially a three-dimensional drawing – is often cited as one of the earliest instances of installation. It came about as Lissitzky's pictorial and graphic work overran its frame. These relief compositions, protruding into space, absorb the room.[26] Lissitzky described his Proun works as an 'interchange station between painting

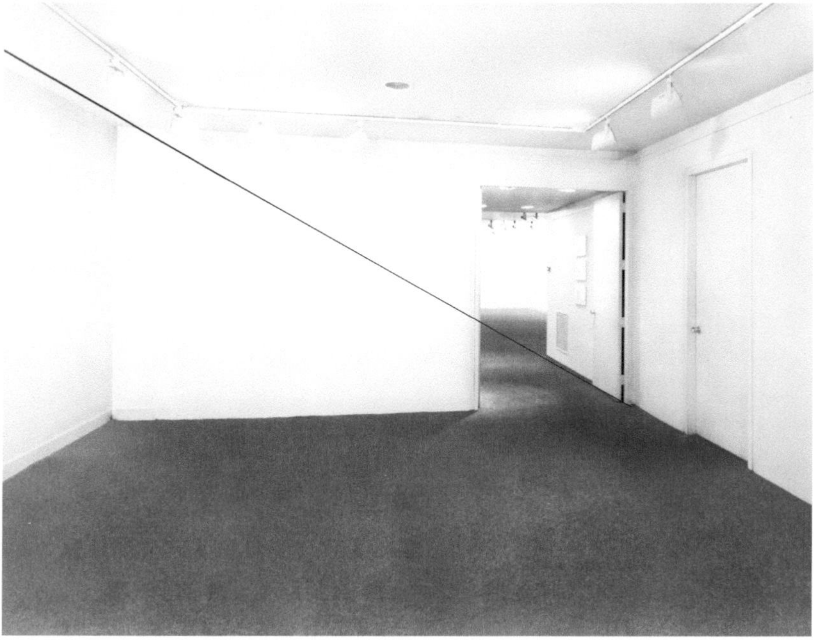

Figure 1.7 *Untitled (One of Four Diagonals)*, 1970, Fred Sandback. Black elastic cord. Situational: spatial relationships established by the artist; overall dimensions vary with each installation. Installation view Fred Sandback, Dwan Gallery, New York, 4–30 April 1970. Photograph courtesy of Fred Sandback Archive. © 2024 Fred Sandback Archive.

and architecture'.[27] The viewer was required to be active – moving screens and bending to see parts of the work hung low to the ground – the intention being that the viewer's movement would shift the works' optical effects.[28]

Installation art addresses the viewer as *embodied* and, according to Claire Bishop, 'integral to the completion of the work'.[29] For Bishop, the viewer's physical participation disrupts a singular perspectival view in favour of 'sensory immediacy' which both *activates* and *decentres* the viewer.[30] Installation also diffused objects; Hans Ulrich Obrist writes, 'The constant element [in installation art] is a refusal or a rejection of the limits of objecthood. Similarly, there is a refusal to address a single object without exploring its interactions, its relationships, the interstasis of objects and contexts, not only in space, but also in time.'[31] For both Bishop and Obrist, installations are relational fields comprising viewers, not the objects perceived by them. Sandback envisioned a work 'which became less of a thing-in-itself, more of a diffuse interface between myself, my environment, and others peopling that environment. . . . A drawing that is habitable.'[32] The specific contexts in which installations take place are therefore folded into the works and may be occupied and, hence, resemble architectural sites.[33]

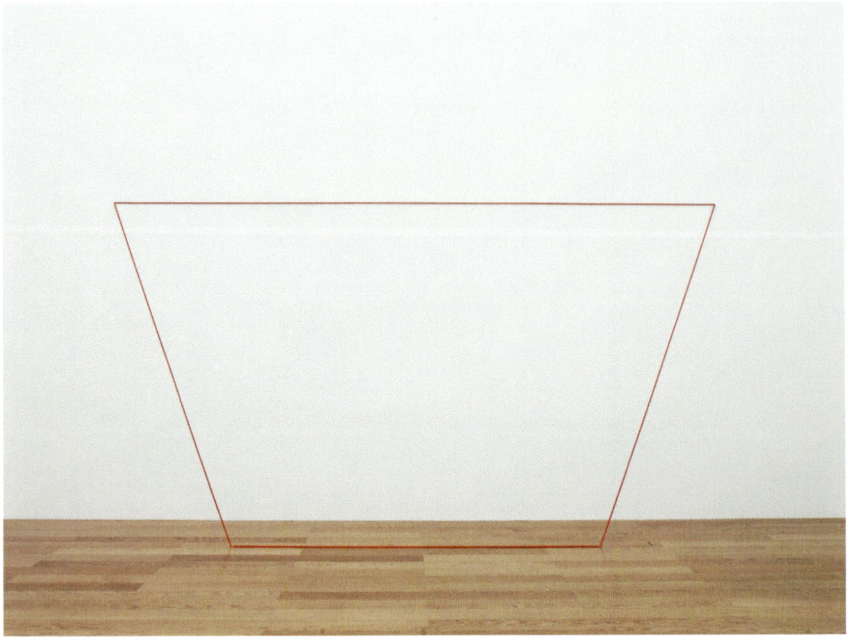

Figure 1.8 *Untitled (Trapezoid)*, 1968, Fred Sandback. Orange acrylic paint on 3/8-in. diameter elastic cord. 88 1/4 × 168 × 24 in. (224.1 × 426.7 × 61 cm). Photograph by Thomas Cugini, Zurich, courtesy of Fred Sandback Archive. © 2024 Fred Sandback Archive.

In *Measurement Room* (1969), Mel Bochner made a series of lines with black tape accompanied by measurements marked in transfer lettering, as if the gallery were a dimensioned architectural drawing.[34] In such drawings, dimensions typically provide supplementary information, directing attention to the thing dimensioned. By annotating the gallery itself, Bochner exploited this effect to produce drawings that efface themselves. While not inviting viewers to delve into imaginary worlds, this occupiable drawing, or 'three-dimensional blueprint' as described by Katharine Stout, revealed its architectural and institutional contexts.[35]

Giving visceral sense to the idea of the drawing-installation that reveals its context, Gordon Matta-Clark's *Splitting* (1974) disclosed the uneven effects of the architectural line. He snapped a chalk-line down the wall of a New Jersey house due to be demolished. He then retraced it with a sabre saw, cutting the house in half. The first line was the work of a moment, the second a substantial labour. With the house supported on jacks, he trimmed the foundations so that half the house could be lowered back on a slight angle, exposing slivered views into an interior that was now uninhabitable. The house was rendered

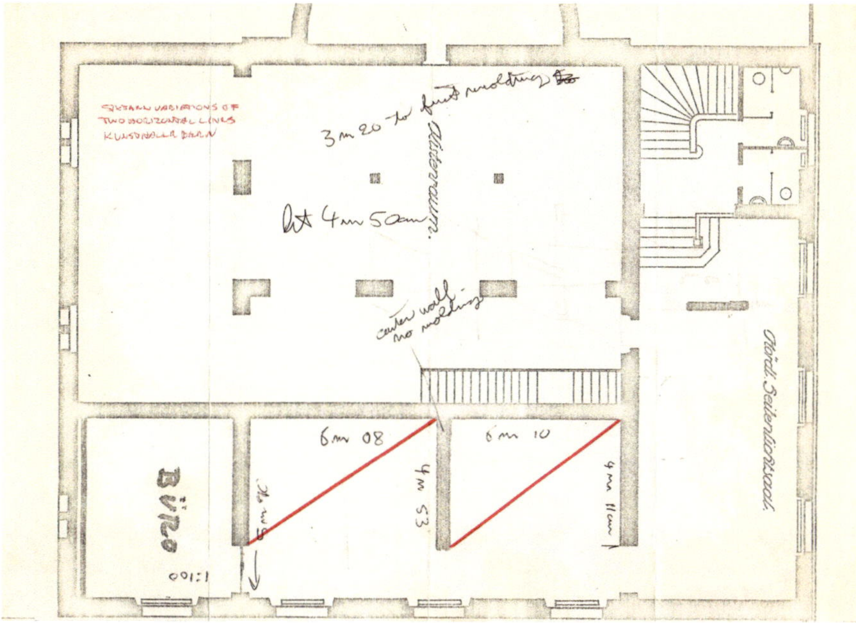

Figure 1.9 Artist's annotated floor plan for his exhibition at the Kunsthalle Bern, 1973, Fred Sandback. © Fred Sandback Archive.

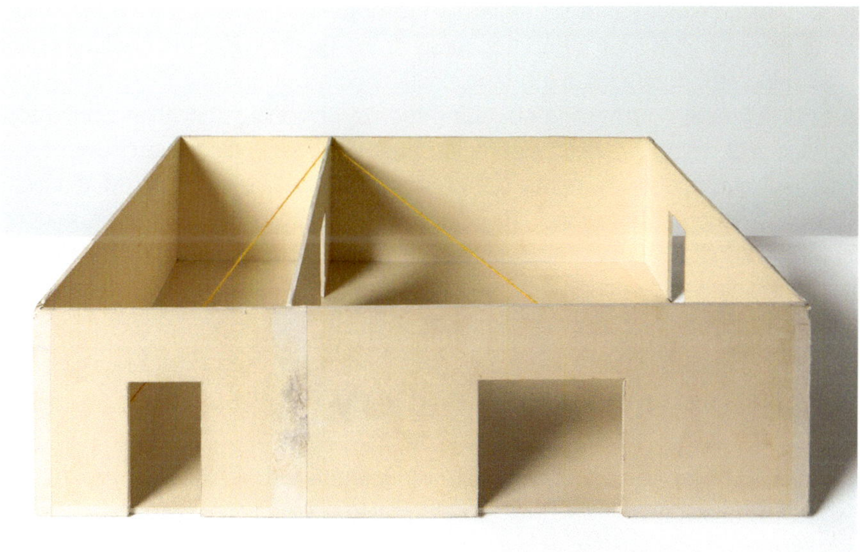

Figure 1.10 Model for Sculpture (Galerie Heiner Friedrich), 1973–4, Fred Sandback. Foamboard and acrylic. 6 19/64 × 21 17/64 × 14 3/8 in. (16 × 54 × 36.5 cm). Kunst Museum Winterthur, purchased by the estate of Curt and Erna Burgauer, 2010. Photograph by Philipp Hitz, SIK-ISEA, Zurich. © Fred Sandback Archive.

precarious, and Matta-Clark took physical risks to manifest this single line. Mark Wigley's parsing of the work emphasizes the stretch of these cuts: beyond the site of the building and into its own documentation as an 'extended intermedia performance'.[36] The photographic record, as exhibited and as included in the artist's book he produced, situates these building cuts as already photographic. This documentation of pieced-together images demonstrates the dominance of the line of the cut, yet with the effect that 'building elements were moving to prop up a line rather than a line being cut through a stationary building'.[37] Alan Moore's review of Matta-Clark's 1974 exhibition likened his penknife – in cutting into and manipulating the photographs – as reiterating the gestures of his chain saw.[38] Wigley writes that 'multiple images – and even multiple sets of multiple images – are necessary to capture the loosening and undoing of an apparently singular stable object'.[39] *Splitting* reveals the disparity between the drawn abstract line and a physically enacted line, 'thick with matter and error and gravity'.[40] In this gap, writes Francesca Hughes, 'Matta-Clark sites himself, his oeuvre, and crucially, his body, his very physicality'.[41] If Bochner's dimensions make the institutional boundaries of the gallery explicit, Matta-Clark's cut reveals the line's potential violence.

It is not a huge conceptual leap from *Splitting* to *9119 St. Cyril*; both are works of demolition and revelatory undoing. Deliberate acts erase the delicate human order of the house: divulging a seam of light in the former and a full-scale plan in the latter. Together, the projects specifically call into question the linearity and ideality of architectural order embodied in drawing. Drawing, by 1980, looked quite different from how it had appeared at the beginning of the century. It was now understood as something temporally enacted, a bodily activity rather than a kind of image. A drawing could be three-dimensional rather than bound to a surface and critically reveal its context rather than being placeless.[42] Lines were not neutral, pictorial elements but concrete and active elements of the built, thought and practised world. The extension of drawing into full-scale material enactments directed attention to the here and now rather than referring to elsewhere and elsewhen.

Architecture in the gallery

Architects, however, cannot dispense with reference so easily. Too much of their work requires representative description; scaled drawing and a well-trained scalar imagination are part of their projective toolkit. By the 1970s, architectural drawing had long been a settled technique. Although modernism and photography had brought about changes in the nature and role of drawings in practice, it was broadly understood that architects produced sets of plans, sections, elevations and details sufficient to guide future construction. In the late 1970s, however, the expanding nature of drawing in art began to have an impact on how architects drew. A series of architectural exhibitions held in New York at this time revealed new ambitions for drawing, linked to questions about the linearity of architectural practice. By the 1990s, a critical re-evaluation of architectural drawing was well underway.

Prior to the 1970s, exhibition held importance for architecture as a way to participate in public dialogue. *Modern Architecture* (MoMA, 1930) was a manifesto for the International Style, while in England, *This Is Tomorrow* (Whitechapel Art Gallery, 1956) featured architects working alongside artists, to give just two examples. Architecture always sat slightly awkwardly in the gallery, featured by drawings, models and photographs, all second-hand to the acts of designing and drawing.[43] Exhibition, however, was to become an increasingly crucial mode of practice for architects towards the end of the twentieth century. Exhibitions could display (or create) genealogies and stylistic unities, and, as Zoë Ryan argues, 'position ideas and present arguments, shed light on current concerns, suggest future directions and draw connections with larger theoretical, political, and cultural conversations'.[44]

After a number of significant public exhibitions of architectural drawings in the 1960s, the Leo Castelli Gallery in New York hosted *Architecture I* (1977) to demonstrate that architectural drawings could be marketable commodities.[45] The show, curated by Pierre Apraxine, included works by eight men: Raimund Abraham (1933–2010), Emilio Ambasz, Richard Meier, Walter Pichler (1936–2012), Aldo Rossi (1931–97), James Stirling (1924–92) and Robert Venturi (1925–2018) working with John Rauch (1930–2022).[46] Never before had such a prominent commercial gallery shown work of this type; other Castelli Gallery artists at this time included Jasper Johns, Roy Lichtenstein (1923–97) and Andy Warhol (1928–87). Although few works actually sold, the show was well-patronized and well-reviewed.[47] More importantly though, as architectural academic Jordan Kauffman argues, *Architecture I* advanced a significant proposition: that architectural drawings were important modes of inquiry and practice, independent of whether a building was their ultimate target.[48]

The show included a mix of speculative, forthcoming and built works. Pichler's pencil and ink drawings were amongst the speculative ones: drawings for his *St. Martin a.d. Raab* project – including *Observatory* (1974) and *Two Rooms* (1975) – show isolated structures, lightly grounded with faint contextual settings: spaces of lonely intimacy and solitude (see Figures 1.11 and 1.12). Abraham's abstract axonometric drawings of his *Nine Houses* project, vividly rendered in coloured pencil, were similarly never intended to be built (see Figure 1.13). Meier and Venturi and Rauch, by contrast, presented models and drawings of buildings from their offices. Stirling's contribution included sketches as well as finalized presentation models. These various drawings were isolated out of their

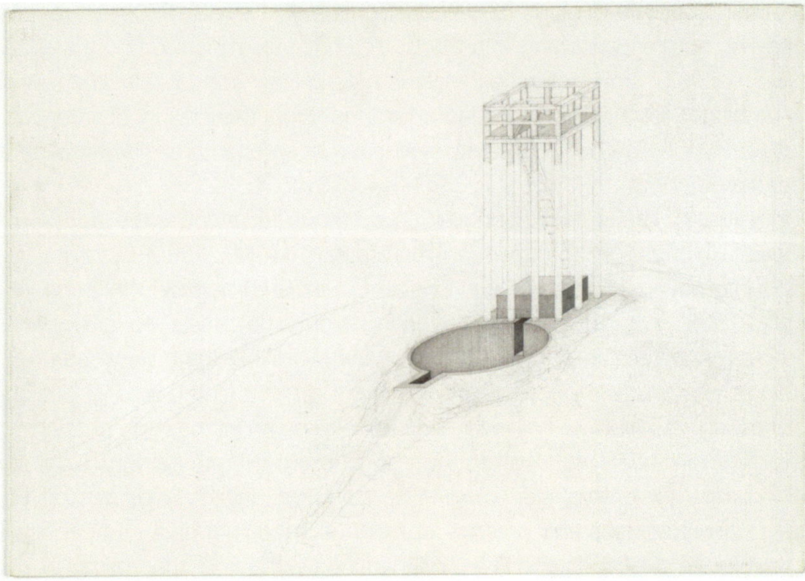

Figure 1.11 *Observatory*, project: perspective, 1974, Walter Pichler. Estate Walter Pichler. Digital image, The Museum of Modern Art, New York/Scala, Florence.

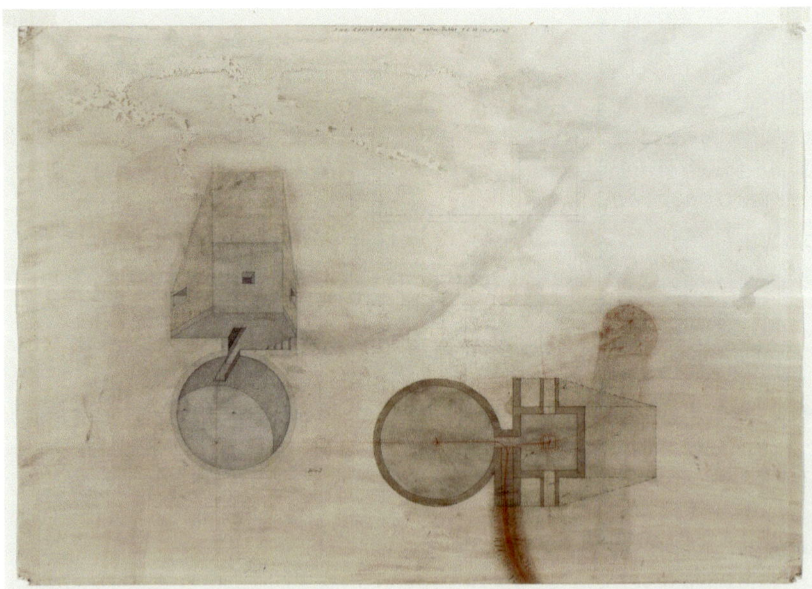

Figure 1.12 *Two Rooms*, project: perspective and plan, 1975, Walter Pichler. Estate Walter Pichler. Digital image, The Museum of Modern Art, New York/Scala, Florence.

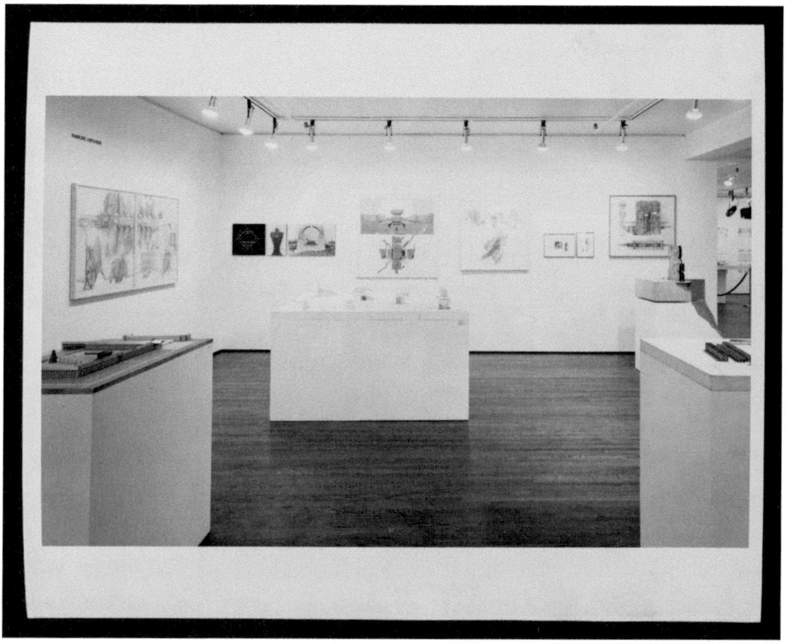

Figure 1.13 *Nine Houses* project, *Architecture I*, Leo Castelli Gallery, 1977, Raimund Abraham. Photograph by Bevan Davies. Leo Castelli Gallery records, Archives of American Art, Smithsonian Institute.

design processes, separated from their referent buildings, framed and matted similarly and hung on the walls.[49] Frames, mats, vitrines, plinths, white walls and spotlights isolated the works from each other, using a conservative spatial language articulating discreteness and separation.[50]

This show contrasts sharply with *Line as Language* which explicitly undermined the distinctness of the art-object. The majority of the works in *Architecture I* were inescapably referential and illusionistic – they directed the viewer to another place. Scale articulated this displacement: viewers were obliged to supply themselves with imaginary diminished bodies to project themselves into these places. Commercial imperatives reinforced the priority of art objects and combined with the underlying logic of the architectural drawing as a miniaturized representation. The idea that architectural drawings could be independently appreciated for their intrinsic qualities rather than as a surrogate for buildings was to be a crucial shift in architecture during the 1970s and 1980s.[51] Kauffman argues that architectural drawings have architectural value in their own right, 'had ontological implications for drawing specifically, and for architecture generally. . . . During this period architectural drawings became the driving force of architectural debate, not for what architects put in them, but for what others asked them to be and saw in them.'[52] The importance of exhibiting grew, closely linked to a new emphasis on speculative practice:

> The whole exhibitionary complex of architecture – exhibition, museums, and archives – has often been understood in this sense as both a symptom and a cause of this new turn towards 'paper architecture'. . . . Unhinged from the political, ideological, and economic forces conditioning architectural production, the exhibition provided a space in which these forces could be displayed, their implications revealed, and their trajectories acted out.[53]

In a follow-up show, *Architecture II: Houses for Sale* (Leo Castelli Gallery, 1980), curated by Barbara Jakobson, architects were asked to design 'visions for the modern house'.[54] The brief specified a plot at least an acre in size, a budget of $US250,000 and invited specific attention to construction and climate. The drawings and models in the show were for sale; in addition, the designs could be purchased as buildable prototypes.[55] The dual nature of the drawings offered the possibility that they might have value in both the circuits of the art world and those of architectural practice. *Line as Language* and the Castelli shows were very different, but in their own ways they each assigned priority to the drawing as a physical artefact.

One effect of this new emphasis was the blurring of model and drawing. Arata Isozaki's (1931–2022) contribution to *Architecture II* was a set of metal relief plates (*House of Nine Squares*), building up the lines of the drawing as edges and giving it depth, and Peter Eisenman integrated the drawings and models of his *House El Even Odd* into a hybrid composition (see Figures 1.14 and 1.15). Abraham was later to write: 'A drawing for me is a model that oscillates between the idea and the physical or built reality of architecture. It is not a step towards this reality and in this respect, it is autonomous.'[56] The distinction between two- and three-dimensional representations diminished in importance compared to the ambiguous role of both in relation to the linear model of practice. Practically, the models of *Architecture I* and *II* were harder to sell than the drawings. Kauffman reasons that their conservative presentation on plinths made them seem dated and their construction from lightweight materials implied a lack of durability.[57] Perhaps also their lack of ambiguity and apparent straightforwardness inhibited them from being read as anything other than representations for building. In 1976, Eisenman curated the exhibition *Idea as Model* (Institute of Architecture and Urban Studies, New York). The accompanying book was published in 1981, and so *Idea as Model* can be seen as bracketing the two Castelli shows. Eisenman's intention was to examine conceptual models as being unpremeditated and generative, precise yet experimental, not 'propaganda for persuading clients, but rather as studies of a hypothesis, a problem, or an idea of architecture'.[58] Although ostensibly a show about models, it was just as much a show about drawing. It

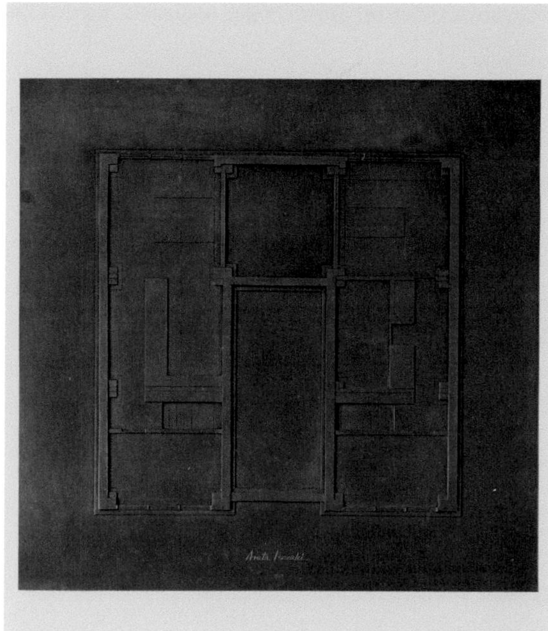

Figure 1.14 *House of Nine Squares*, *Architecture II: Houses for Sale*, Leo Castelli Gallery, 1980, Arata Isozaki. Photograph by Bevan Davies. Leo Castelli Gallery records, Archives of American Art, Smithsonian Institute.

incorporated a mix of models, dioramas, reliefs, installations, prototypes, toys, sculptures and miniatures, many of which relied on a tension between the two-dimensional and three-dimensional.

Rafael Moneo's *A Study of a Wall with Windows* presented a triptych: a photograph showing the wall from a recently completed building, a drawing of the same wall in axonometric and a cardboard model in shallow relief, inviting viewers to consider the differences and implying the model is a step further from building than the drawing, rather than closer to it.[59] The work aimed 'to avoid the problems of change in scale by admitting the possibility of materializing the drawing into three dimensions . . . reversing the standard operation of reducing a three-dimensional object to a two-dimensional representation'.[60] John Hejduk (1929–2000) also used shallow relief in *Natural History*; his later work, *The Thirteen Watchtowers of Cannaregio* (1974–9), continues this (see Figure 1.16). Charles Moore (1925–93) and Richard Oliver presented 'vacu-drawings', vacuum-formed reliefs made for a blind client.[61] Bernhard Leitner contributed *Four Spatial Transformations of a Grid*, a Perspex model for one of his sound-space installations[62] (see Figure 1.17). He drew diagrams of lines and dots on layered sheets of acrylic to give them three-dimensional depth. A figurine

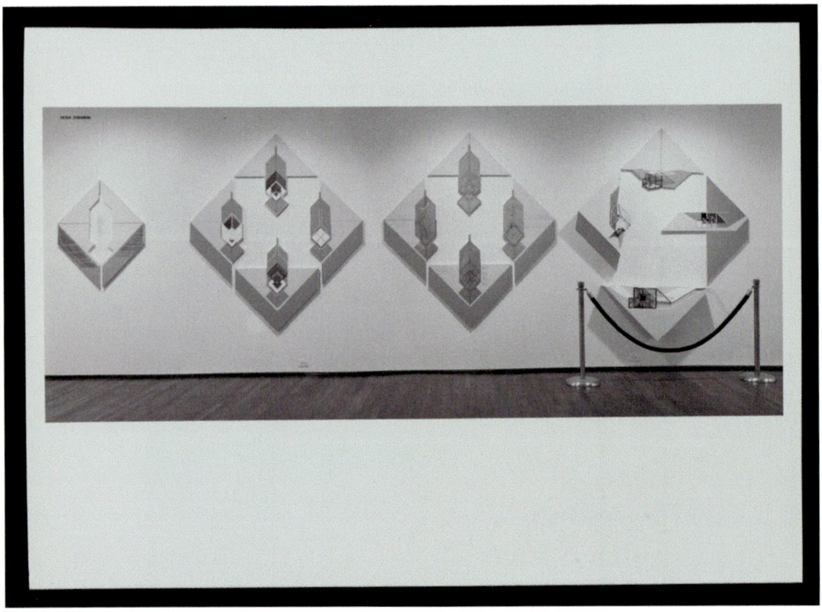

Figure 1.15 *House El Even Odd*, *Architecture II: Houses for Sale*, Leo Castelli Gallery, 1980, Peter Eisenman. Photograph by Bevan Davies. Leo Castelli Gallery records, Archives of American Art, Smithsonian Institute.

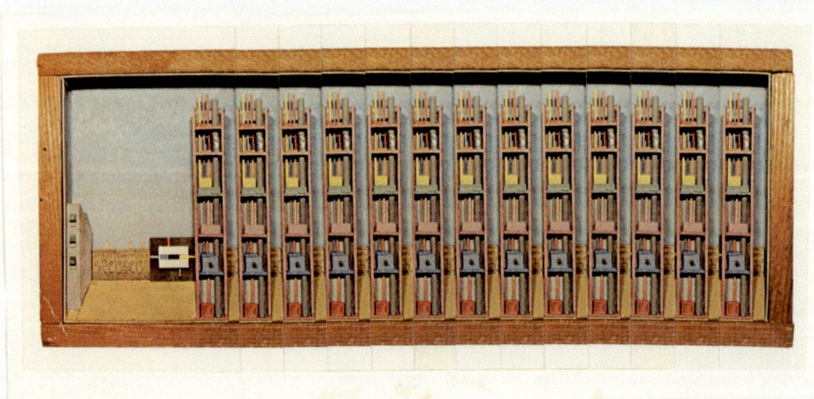

Figure 1.16 *The Thirteen Watchtowers of Cannaregio*: section comprising views of a presentation model 1974–9, John Hejduk; paper collage of reprographic copies; 6 3/8 × 12 5/8 in (16 × 32 cm); DR1998:0093:002. John Hejduk fonds, Canadian Centre for Architecture © CCA.

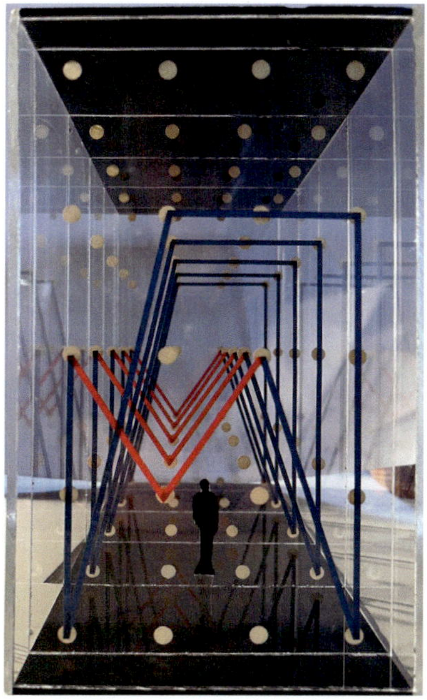

Figure 1.17 *Variation of a Grid/Four Spatial Transformations of a Grid*, 1977, Bernhard Leitner. Atelier Leitner, Vienna.

indicated scale and invited viewers to imaginatively enter this spatial diagram. Included in the book, but not the show, Eisenman's famous anamorphic model for *House X* appeared as an axonometric from a certain point of view but bizarrely skewed from any other viewpoint.

The 1981 book was not exactly a catalogue. It included more works than shown in 1976 and notably highlighted what a number of participants had come to see as the show's failures. Richard Pommer contributed two essays: in one, 'Postscript to a Post-Mortem', he writes that *Idea as Model* expressed nostalgic escapism; in the other, 'The Idea of "Idea as Model"', he considered the show a 'recoil from material reality' into allusion.[63] Christian Hubert wrote that the model, as a scaled miniature, is invisibly framed: 'the space of the model . . . demarcates a limit between the work and what lies beyond'.[64] Allusion and displacement are two sides of the same coin. Although contributors to the show and catalogue made reference to artists, the prevailing sense was that the exhibition stayed within disciplinary bounds. In spite of its apparently avant-garde aims, the show took a fundamentally defensive posture: to reaffirm architecture's specific disciplinarity. It was disappointing, Hubert noted, that Matta-Clark's contribution was withdrawn at the last minute.[65] This absence notwithstanding, his contribution has since become legendary. Having initially proposed to intervene in the gallery by cutting into the walls, he arrived on the afternoon of the show's opening with a BB gun, proposing to shoot out some of the gallery windows and use them as frames to display photographs of housing projects in the Bronx. Art historian Pamela M. Lee describes that he then 'proceeded to shoot out all the windows on the floor of the Institute, all the while raging against its esteemed members and the architectural ideologies they supported'.[66] He was kicked out of the show and the windows boarded up before the opening. Aggressive and deliberately offensive, Matta-Clark's actions also exposed the implicit constraints of the undertaking: his actions drew attention to the sense of isolation, withdrawal and limits implicit in the show.

Regardless of whether *Idea as Model* or the first two *Architecture* shows are deemed to have succeeded in their aims, several things are clear. Firstly, the gallery had emerged as a crucial, if still complex site for architecture. Secondly, models and drawings were argued to be architectural inquiries, thoughts or fictions, rather than representations. They could be about, from, in response to or parallel to building, rather than directly enabling its construction. And thirdly, the distinction between two- and three-dimensional modes was less important than the distinction between representation and direct presentation, speculation and feasibility.

One to one

Our study is particularly concerned with two interwoven strands of drawing emerging from the 1970s: in art, the extension of drawing as a transdisciplinary concern with the relationship between ideality and material specificity; and in architecture, the exhibition of hybrid drawing-models with a value independent of their direct role towards construction. Over the next two decades, these strands wove a new framework for understanding architectural representation. History, theory and criticism of drawing flourished, demonstrated in publishing: Robin Evans's (1944–93) *The Projective Cast* (1995) and *Translations from Drawing to Building* (1997) argued that the problems of drawing – its ambiguities and inconsistencies – were a locus of architectural invention; Hubert Damisch's (1928–2017) *The Origin of Perspective* (1994) and Alberto Pérez-Gómez and Louise Pelletier's *Architectural Representation and the Perspective Hinge* (1997) demonstrated the deeply philosophical nature of technical drawing procedures; Catherine Ingraham's *Architecture and the Burdens of Linearity* (1998) brought feminist and psychoanalytic insights to bear; and Stan Allen's *Practice: Architecture, Technique, Representation* (2000) argued that pure theories were inadequate to grasp the complex, paradoxical and impure fields of activity that make up architectural practice. The overriding theme of the many books and articles written on architectural drawing in the 1990s is that transparent representation is both impossible and inescapable for architecture.

An installation from 1997 by artist Glen Seator (1956–2002) shows how this conjunction of art and architecture raised questions of scaled representation. *B.D.O.*, at the Whitney Museum, was somewhere between full-scale model and found object. Seator meticulously salvaged an office interior from the museum that was to be demolished for renovations. The teak-panelled interior, fittings and furnishings, designed by Marcel Breuer (1902–81), were stripped out and reconstructed in a timber-framed box. The entire box was then set at a vertiginous 30-degree angle in the gallery (see Figure 1.18). Viewers looked in through the formerly exterior window. Christian Hubert writes, 'The piece both belonged there and belonged nowhere. It turned the model's desire for physical actuality into a kind of anxious conundrum, and the vertigo one felt looking into the tilted room so delicately poised was matched by its conceptual instability at the dividing line between the worlds of art and architecture.'[67] The installation was not quite the 'real' office, since it had been reconstructed and was no longer habitable; but nor was it strictly a model, being a literal transposition of the physical interior. It reproduced the discomforting, invisible limit characteristic of models. Although it was full-size and made of the original

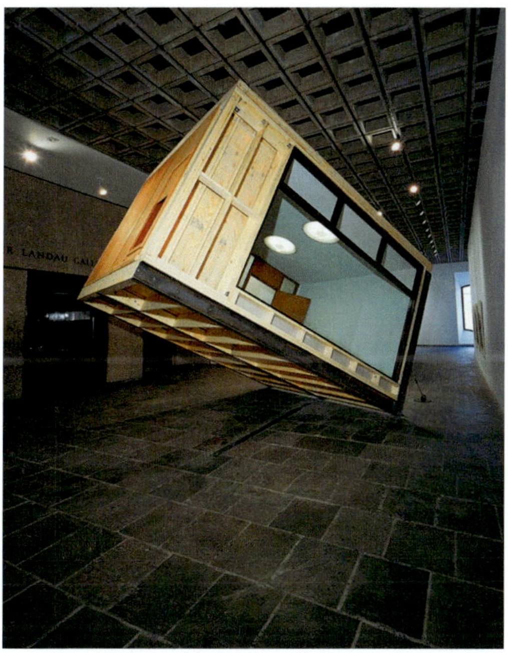

Figure 1.18 Installation view of 1997 Biennial Exhibition (Whitney Museum of American Art, 20 March–15 June 1997), Glen Seator, *B.D.O.*, 1997. Purchased with funds from the Contemporary Painting and Sculpture Committee. © Glen Seator Estate. Photograph by Geoffrey Clements.

materials of the interior, *B.D.O.* became a model of itself, observable but not occupiable.[68] Hubert's sense of vertigo might be less a matter of disciplinary ambiguity (is it art or architecture?) and more to do with this tension between distancing and presencing in the work.

Writing of the 1986 reconstruction of Mies van der Rohe's Barcelona Pavilion, Jonathan Hill (1958–2023) points out the difference between a reconstruction and its original. While the latter is used, the former is an artistic object of reference and contemplation: 'The contemplation of art is a form of visual awareness, of a single object by a single viewer, in which sound, smell and touch are as far as possible eradicated. Placed in a hermetic enclosure and protected against decay, the artwork is seen but never used. The viewer leaves no trace or mark.'[69] As visitors we may walk inside these full-scale spaces, but they resist inhabitation; they remain *model-like*. The full-scale model, even when we can occupy it, is an invitation to suspend the presumed immediacy of our experience. The space we occupy both *is* and *is not* real. In *B.D.O.*, the act of relocating and tipping the Director's office deliberately opens up this disjunction. Formerly, this room was behind the scenes, part of the gallery infrastructure, but now we encounter it as an object. The aloofness (and, potentially, precarity) of the Director – their distance from the viewer – is emphasized at the same time the visceral gravity of the installation's material state is dramatized. The building is restaged within itself. Rachel Whiteread's *Ghost* (1990) and *House* (1993) had previously presented a similar tension: the inversion produced by casting keeps the viewer on the outside.

Simultaneous distancing and presencing is a hallmark of the hybrid full-scale drawing-installation. Melissa Gould's *Floor Plan* (1991), a site-specific work in Linz, Austria, traced in light the plan of an absent building in its former location[70] (Figure 1.19). One hundred and ten fluorescent tubes, flush with the ground, outlined the full-scale plan of a nineteenth-century Reform synagogue in Berlin which had been damaged on Kristallnacht and by bombing in the Second World War and was later demolished.[71] A sound work by Alvin Curran, titled *Notes from Underground*, accompanied the installation. Recordings of singing voices emanated from sixty-four loudspeakers installed in surrounding trenches, as though from under the earth. Victoria Milne writes of the work: 'A blueprint of a building underway is an optimistic promise: the lines mark spaces for new activity and inhabitants. On the other hand, the plans of a place that no longer exists mark only loss; not only of bustle, but sometimes of life, culture, and even promise itself.'[72]

Visitors to *Floor Plan* experience a full-scale rendering of a volume no longer extant; Gould describes this as 'a phantom building, at once present and absent, real and surreal'.[73] The ability for visitors to walk inside the plan of this building, most evocatively between dusk and dawn, undoes this sense of loss Milne refers to: the building is inhabited once again, differently. *Floor Plan* is not a plan drawn

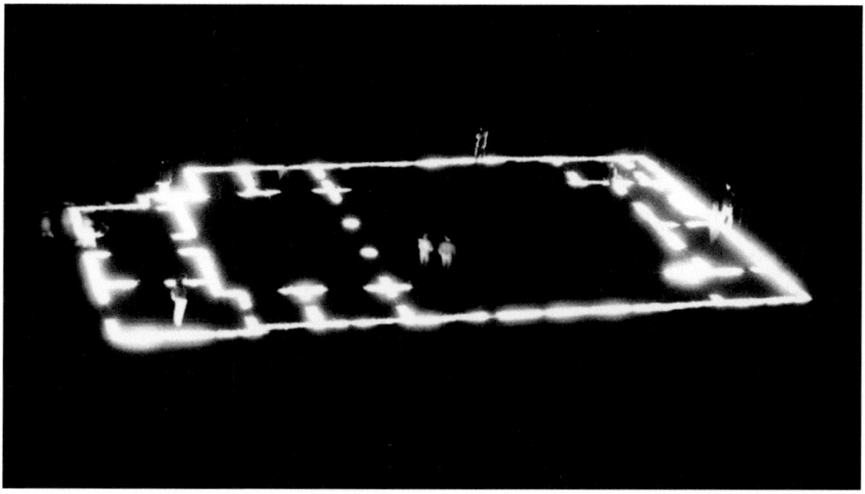

Figure 1.19 *Floor Plan*, a Holocaust Memorial installation at the 1991 Ars Electronica Festival, Linz, Austria, Melissa Gould (MeGo), 57 × 80 ft, 110 fluorescent light tubes. More information @ www.megophone.com/floorplan.html. © 1991, Melissa Gould (MeGo).

as a graphic on a substrate; it is inscribed and materially enacted. At the same time as the viewer is immersed in the effects of lighting and overlaid into the plan, their sense of dislocation is heightened. The full-scale drawing maps onto the body of the user but insists on what is not present. As Bishop writes, such work

> admits a flood of conflicting appeals to the viewer's authentic experience of 'heightened consciousness' (of body, self, place, time, social group) as – paradoxically – *both* an assertion of *and* a decentring of subjectivity. This is because installation art plays on an ambiguity between two types of subject: the *literal* viewer who steps into the work, and an abstract, philosophical *model* of the subject that is postulated by the way in which the work structures this encounter.[74]

The dislocating effect of occupying a spatial drawing activates parallel dislocations of memory and loss; the immediacy of place and time are suspended. In *Flattened Room* (1991) Jeanine Centuori, then a student at Cranbrook, used latex rubber to take thin surface-castings of a table setting that were unwrapped as full-scale drawings (see Figures 1.20 and 1.21). Recalling the stripped interior of Seator's *B.D.O.*, a commonplace domestic setting was understood as surface that could be peeled from its laminate. Cut, stretched and mounted like the skin of an animal, the interior formed a graphic figure: 'It is as if the flattened skins were waiting for a breath of volume to peel themselves off the panel and be reconfigured, their

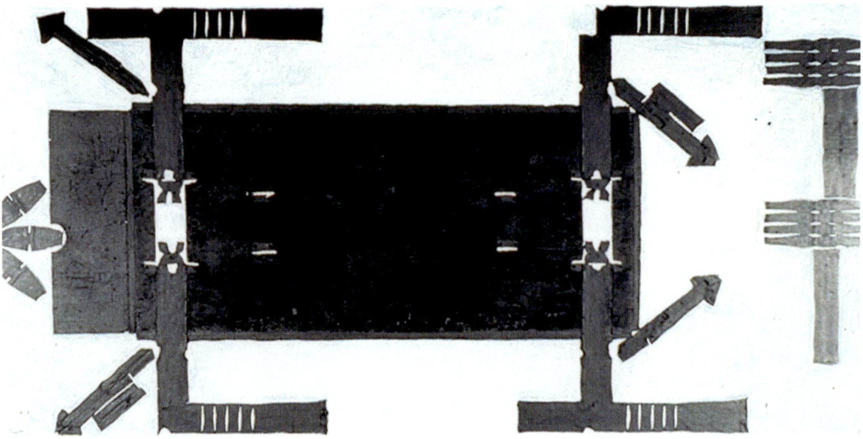

Figure 1.20 *Flattened Room*, 1991, Jeanine Centuori. Latex and paint.

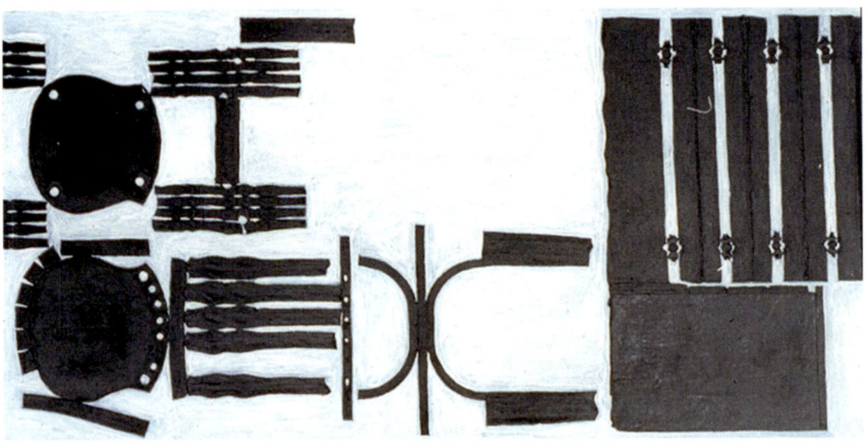

Figure 1.21 *Flattened Room*, 1991, Jeanine Centuori. Latex and paint.

surfaces carrying the memory of the original volumes. We sense their potential to occupy volumetric space and unconsciously fill them.'[75] Reoriented to the wall and reorganized according to the logic of unfolding and the dimensions of the wooden panels to which they are mounted, the surfaces do not present themselves in the relationships we would commonly expect to find them. The viewer sees all sides at once: the knife-scored tabletop as well as its 'splintery' underside and each unrolled face of the table-legs.[76] *Flattened Room* draws the viewer close but defamiliarizes a familiar scene. The cast surfaces picked up traces of material detail and the irregularities introduced by use, emphasizing surface rather than form. As a direct imprint of the table setting, the table is one-to-one: not a symbolic representation but an indexical one.

Cathcart, Fantauzzi and Van Elslander, who worked on *9119 St. Cyril*, went on to produce a series of works in a similar vein. In *Slice* (1999), a rectangular section of gallery wall was cut and rotated 90 degrees around a vertical axis, linking two previously separate spaces. In *Excavation* (1992), a van was cemented upside-down in its parking space and then excavated to reveal its form cast into the ground. In most of their work, only ad-hoc sketches were made prior to beginning. Kent Kleinman wrote of their 'elimination of drawing', with the result that 'there are no means for articulating, judging, predicting, or composing in advance of building'.[77] Another perspective is to recognize the work itself was already a kind of performative drawing, or rather an exploration of latent drawing. Kleinman continued:

> Its territory is already prepared, its site already determined by prior architectural acts. The pit for the buried car in *Excavation*, for example, is precisely the extent of the existing parking spot painted on the asphalt surface; this demarcation, the received site of action, was predrawn. . . . Limits and lineaments set by preceding practices are essential to much of the work.[78]

9119 St. Cyril and *Flattened Room* were drawings *from* building, not towards it. They were situated and site-specific as opposed to off-site and studio-based; they were direct and full-scale rather than mediated and scaled. They point towards drawing that no longer has building as its sole endpoint and disrupts the seeming transparency of architectural lines. These one-to-one embodied enactments of drawing-modelling envision new ways to draw in service of conventional modes of architectural practice and also point towards an expanded sense of such a practice.

Expanded practice

Whether or not models and drawings have found the kind of autonomy to which Eisenman aspired through *Idea as Model*, the period from 1970 to 2000 marked the rise of exhibiting as an important mode of architectural practice and inquiry. It signified the beginning of a fertile interchange between drawings and models. Installation-drawings addressed bodies (notably the bodies of viewer and drawer) at full scale. The crucial quality of these drawing-model-installations was their *ambiguity*, their ability to both place and displace the viewer, trace and escape from the drawer. *Line as Language* pointed to ways to understand architectural drawing that eluded these architectural shows. It complicated the relatively stable body of architectural drawing techniques and conventions. We identify the critical legacies of this period as the loosening of disciplinary assumptions, the open question of how drawings act in space and of how bodies encounter these works. *9119 St. Cyril* has served as an initial prompt to focus our attention on these shifts and the way art practices of the 1970s introduced new questions, themes and practices to architecture. As a school of architecture specifically premised on such transdisciplinary activity, Cranbrook exemplifies this vital exchange but was by no means the only place it occurred. Geographically and temporally, the scope of the examples gathered in this chapter is narrow – mainly North America (particularly New York) in the second half of the twentieth century. Culturally, they sit within traditions of art and architecture arising from the Renaissance in western Europe. There are of course many other contexts in which drawing, modelling, installation and exhibition come into contact outside this lens, and cultures in which these distinctions may not be pertinent.[79]

Drawing features in this experimentation as a situated, bodily activity. Drawings could be three-dimensional or temporal. A line could be a fold in carbon paper, the edge of an existing wall, a chalk-mark designating a cut, a latex seam to be flattened or an annotation. Their physical materiality and their relationship to the bodies of those who made and viewed them became activated. They were no longer understood as passive representations but scrutinized for their active agency. For architects, the early signs of interest in these new ways to draw manifested in the hybridization of drawings and models and new hunches about the exhibition as an architectural site. New critical understandings of architectural representation in the 1990s have been obscured by the rapid takeover of architecture by digital tools and techniques. It is perhaps easier to tell the story of a technological paradigm shift than to articulate a complex, layered and shifting landscape of concepts and practices. In the next chapter we broaden our scope, to consider how computation featured in this landscape, positing a relationship between computational thinking and full-scale drawing. In our view, computation has not suspended or sidestepped questions of scale and embodiment but made them more urgent. Matters of technique – the marking of points, the tracing of lines through space and the production of immersive effects – continue to enable speculation, offering potential for new acts of tracing and marking, revealing and recording, immersion and inhabitation, enactment and displacement.

2
DRAWING OUT
SCALE AND THE DISPLACED BODY

Smutty daubings

In *Euclid's Elements of Geometry*, we find the venerable definition that 'a point is that of which there is no part'.[1] Indivisible, infinitesimal and capable of bearing no attributes, the geometric point is pure location without dimension. It *says* nothing more than it *is* – in principle at least – and so demonstrates a perfect correspondence between designation and that which is designated. However, in 1665, when natural philosopher Robert Hooke (1635–1703) peered through his newly invented microscope at some actual printed points, he was fascinated to find them far from what Euclid envisioned:

> I found those points that had been made by a *Copper-plate*, and Roll-press, to be as misshapen as those which had been made with *Types*, the most curious and smothly *engraven strokes* and *points*, looking but as so many *furrows* and *holes*, and their *printed impressions*, but like *smutty daubings* on a matt or uneven floor with a blunt extinguisht brand or stick's ends. And as for *points* made with a *pen* they were much *more rugged* and *deformed*.[2]

The microscope revealed the finest marks, made by the most precise instruments, to be like crude markings scored into rough ground with a burned stick (see Figure 2.1). The idea of points as '*simple* and *uncompounded bodies*'[3] suddenly seemed ludicrously abstract. Drawn points disclosed minute spaces. In reverie,

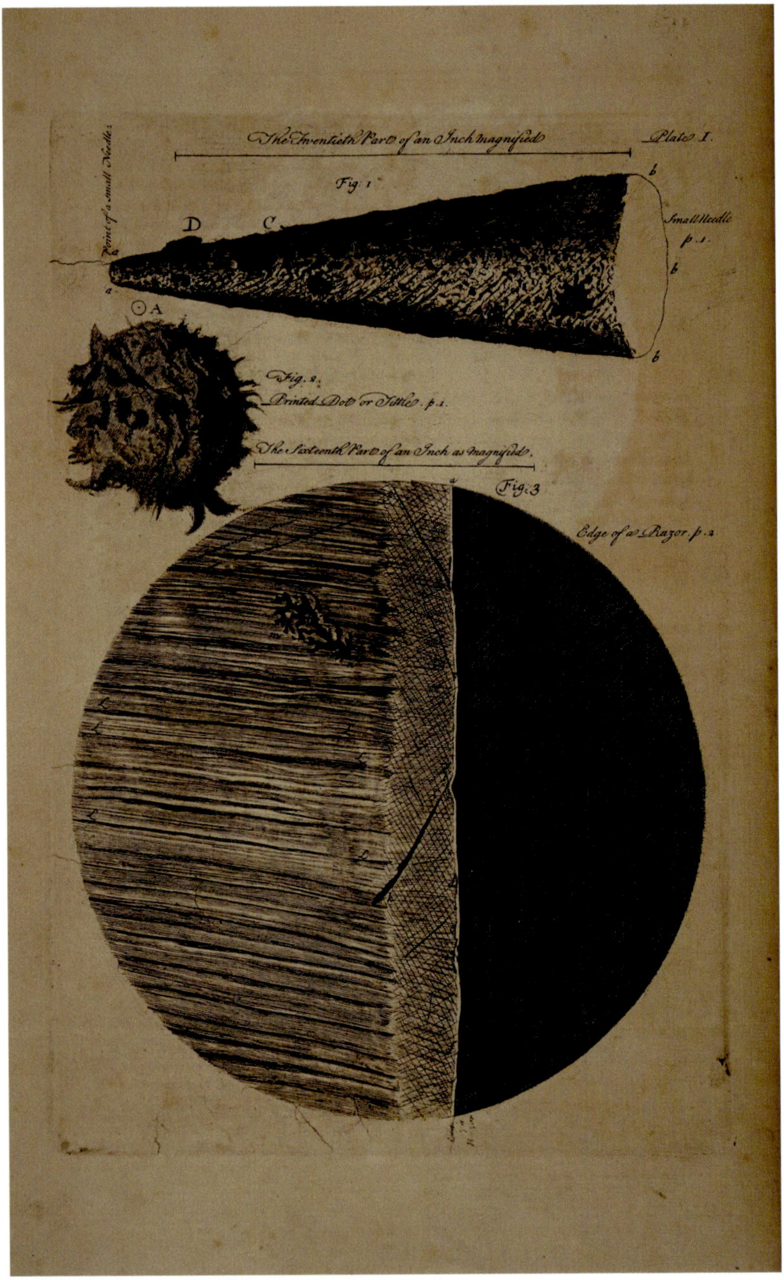

Figure 2.1 Microscopic view of a needle ink blot and razor edge, from *Micrographia*, 1665, Robert Hooke. Courtesy of Science History Institute.

Hooke imagined the apparently smooth paper as a landscape and speculated the earth and moon themselves might be mere points when viewed from far enough away.[4] Euclid's world is free from matter, apprehended and manipulated by way of rational axioms and diagrams. Hooke's world is inescapably material, present to the body and entangled with material imagination.

In the difference between Euclid's notional points and Hooke's smutty daubings we find, in miniature, the tension between rational and embodied modes and between the transparency of drawing and its opacity. The divergence between the cognitive and corporeal sides of drawing exposes another iteration of the Cartesian split between mind and body.[5] Drawing, however, can never be entirely one or the other; rational and embodied modes are not entirely discrete. Rather, they are in a productive dialogue with each other. In fact, as this chapter suggests, dematerialized rational modes of drawing involve intensive embodied activity. Precision, transparency and abstraction are not purely technical matters but imaginative and ideological ones. In her study of accuracy and error in architecture, Francesca Hughes refers to 'the fetishization of precision' that is 'more to do with an aestheticization of control', exemplified by the absurd accuracy of digital tools that enforce 'a degree of precision that is always-already redundant to any process of materialization it might engage in'.[6] In the physical world, precision cannot be assumed but must be achieved laboriously. In other words, considerable physical activity is needed to make Euclid's and Hooke's worlds touch.

Computers seem to realize Euclid's universe at the expense of Hooke's. Emmons argues many uses of computers dissipate embodied disciplinary knowledge and diminish architectural imagination, to the extent that they encourage us to treat drawings as rational, transparent proxies rather than material scenes of embodied imagination.[7] But computation and digitization begin long before electronic computers. The idea of the calculable drawing arises well before the computer and was in fact one of the conditions for its arrival.[8] Three-dimensional computation depends on the idea that the world can be modelled geometrically. In this chapter, we look back to early techniques for converting the material world into geometry: perspective, scanning and surveying. These are not only precursors to electronic computation but also instances of embodied, spatial drawing. We trace these methods forward to contemporary lidar scanning and the paradigm of touchscreen computing. A scalar imagination comes into play in questions of precision. Euclid's point is not really related to the material world in terms of size; it is ideal, not small. The Euclidean definition of a point as pure location without dimension is fundamental to measured drawing and digital space. But how can such abstract points be physically located? Marking them exactly implicates a host of instruments and techniques: pins, pens, pegs, scribes, styluses, threads, crosshairs, compasses, probes, lenses and lasers make visible the embodied labour of exactitude. This labour is often focused on sharpening things, aligning them finely and minimizing variation.

In keeping with our emphasis on drawing as an *activity* rather than an artefact (i.e. drawing as verb not noun), we approach perspective, scanning and surveying not as means to make two-dimensional drawings but as themselves three-dimensional *acts of drawing*. Accordingly, we blur the distinction between a drawing and the apparatus or setup used for drawing. In the previous chapter, drawing expanded in the twentieth century and so raised questions for architecture. In this chapter, we project this understanding of drawing backwards, on the hunch that drawing contained the seeds of its own expansion. Perspective drawing, scanning and surveying (in both their early and contemporary forms) establish systematic relationships between the space of the drawing and the space where the drawing takes place. A representational paradigm sees drawing to be mainly concerned with reflection or correspondence; here, we explore the way it constructs an *interface* between an unruly material world and a pristine rational one. Points, in the form of tick-marks, crosses, indents, pinpricks and mouse-clicks, record a drawing's construction. As much as points thrive on measurement, they belong to both embodied and rational spaces simultaneously as a kind of crossing or contact between them. These points become places of passage. Because points are dimensionless and specified as pure positions, they serve as transfer-points, crossings or spot-welds connecting pictorial space with the real world.

Perspectival apparatuses

Perspective drawing is a rational, technical procedure that defines and manipulates relationships between points. A scene to be drawn is invented as a configuration of linked points, organized by two particularly crucial ones: the viewpoint and the vanishing point. The image registers on a picture plane, an abstract geometric plane that is also the literal surface on which the drawing is being made. Lines are imagined to pass through the picture plane, so that the virtual space on the far side and the space in which the drawer moves are made continuous. By accounting for the physical presence of a viewer, perspective images can appear as if the surface offers a transparent view into another space. For Leon Battista Alberti (1404–72), painting began with a mental trick: 'First of all, on the surface on which I am going to paint, I draw a rectangle of whatever size I want, which I regard as an open window through which the subject to be painted is seen'[9] (see Figure 2.2). The drawing is imagined as a window or mirror to enable the scene to be viewed directly; this frame opens to another space and organizes its composition. This effect of transparency, however, is difficult

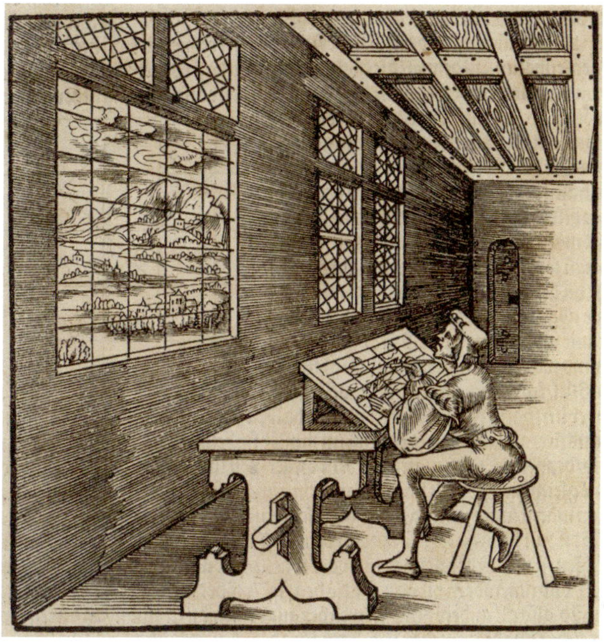

Figure 2.2 *Draftsman at the Table*, Hieronymus Rodler; Johann II, Prince, Count Palatine, 1531. New York Public Library Digital Collections.

to achieve. Perspective was, at first, a set of corrections and approximations to create the visual impression of depth in an image. Beginning with things like tiled floors and vertical walls, it encompassed progressively more difficult components: colonnades, solid forms, curved and angled objects, reflections, rainbows, clouds and human bodies.[10] Once these methods became integrated into a consistent system with orthographic drawing, they were no longer seen as separate technical practices, but as a consistent, unified *space* of drawing. To put physical things into this space, they must be reduced to a set of related points. A brick, for instance, could be drawn taking just eight vertices, assuming that its rounded and chipped edges could be discounted as irrelevant. A more complex or detailed object would require more points to define it. The three-dimensional brick converted to two dimensions is thus *more approximate*, but at the same time *more precise*. It can be transformed virtually, positioned and manipulated in ways real bricks cannot. The Euclidean brick takes the place of the Hookean lump of fire-baked clay.

Practically, this conversion involved the drafter marking point after point, plotting intersecting lines, transferring measurements from real to drawn space using dividers, projecting lines, fixing points in space, cross-referencing between views and connecting dots on a drawing board pricked full of holes. In the early history of perspective technique, there are many depictions of how to construct these drawings. Sometimes these are technical diagrams, abstract constructions of lines and points. Often, they contain a rendered object – a candlestick, column, archway – that is embedded in these lines to convey the idea of a real object that is being cast into perspective. There are also several images which show perspective being constructed using elaborate (sometimes obscure) instruments: frames with grids of wires, threads, mirrors, fixed viewfinders and pointers. In one such image from Albrecht Dürer's (1471–1528) *Treatise on Measurement* (1525), two men work together to set a lute in perspective (see Figure 2.3). A cord, weighted at one end to maintain tension, passes through an eyelet mounted on the wall. The other end stretches through a rectangular frame and is fixed to a stylus touching the lute's body. A crosshair is positioned in the frame, locating the point where the cord passes through it. When the stylus and cord are passed back through the frame, a hinged panel (the picture plane) can be swung into place and the point pricked onto it. The lute is seen coming into view as these sampled points accumulate through a laborious embroidery.[11] The fewer the points needed to describe the lute, the more quickly it could be produced. The lute is sampled intelligently; the drawers judge which contours to trace and how much detail is needed to make the picture. Four eyes move around the apparatus, attending to details, squinting to line things up, glancing

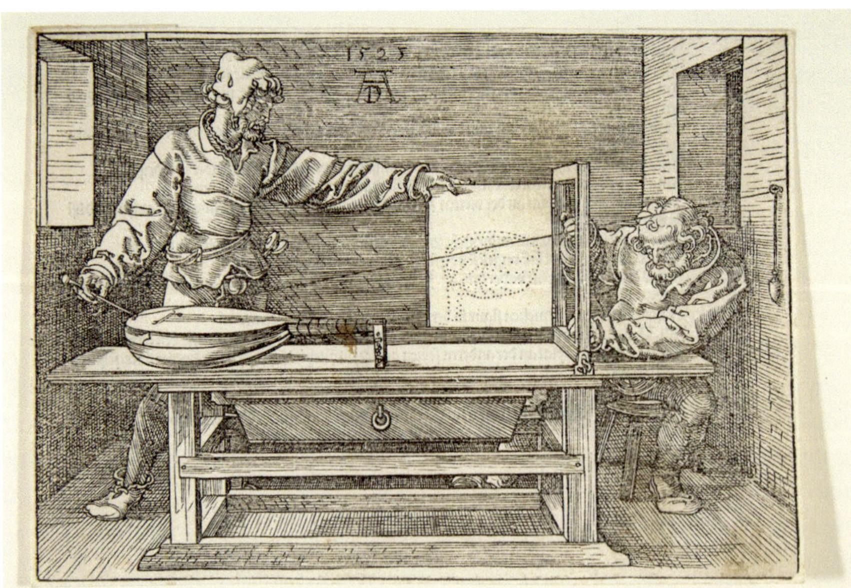

Figure 2.3 *The Draughtsman of the Lute*, Albrecht Dürer, n.d. [*c.* 1525]. Woodcut. Gift of Henry Walters, 1917. The Metropolitan Museum of Art Collection, New York.

at one another, perhaps pausing to look out the window. The virtual eye of the perspective, however, is screwed into the wall, only able to see through an intricate embodied performance in which the body is mechanized and vision imagined materially.

It is unclear whether Dürer recommended this as a literal method to make perspective images, or whether the engraving records a didactic demonstration, or whether it is an elaborate hypothetical.[12] In any case, Dürer supplies his readers with a mental picture and engages their bodily imagination to help them make sense of the abstract technical operations needed for perspective. In another illustration from the same book, an eyepiece on an adjustable rod constrains the drawer's eye while he marks points on a translucent picture plane (see Figure 2.4). The graphic history of perspective methods is filled with such images which show perspective construction as a three-dimensional drawing sharing the space of the drawer (see Figure 2.5). Whether literal or imagined, they recall Fred Sandback's corded spatial drawings (see Chapter 1). Physical objects (a candle, a table, a person) are rendered as three-dimensional bodies touched by geometric lines that connect them to each other and through the picture plane. These temporary, full-scale, situated drawings are a means to negotiate passage into a rational geometric world. This passage harnesses drawing's capacity to

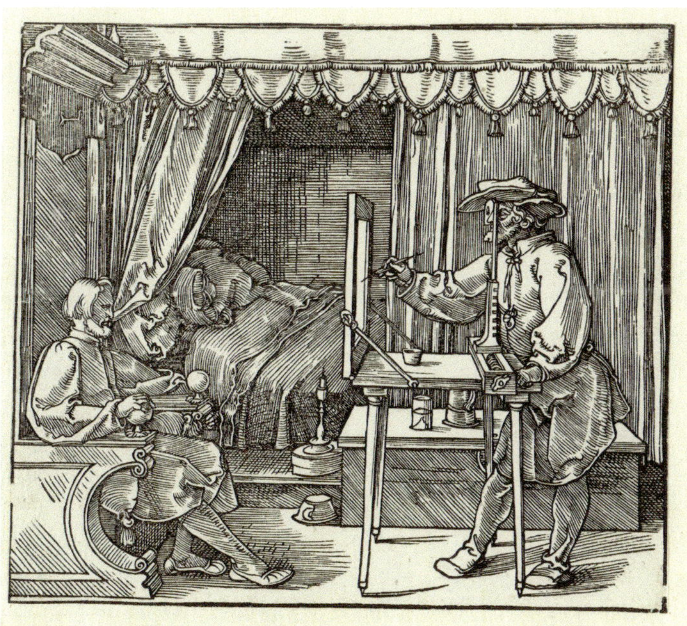

Figure 2.4 *A Draughtsman Drawing a Portrait*, Albrecht Dürer, 1532. Woodcut. Wikimedia Commons.

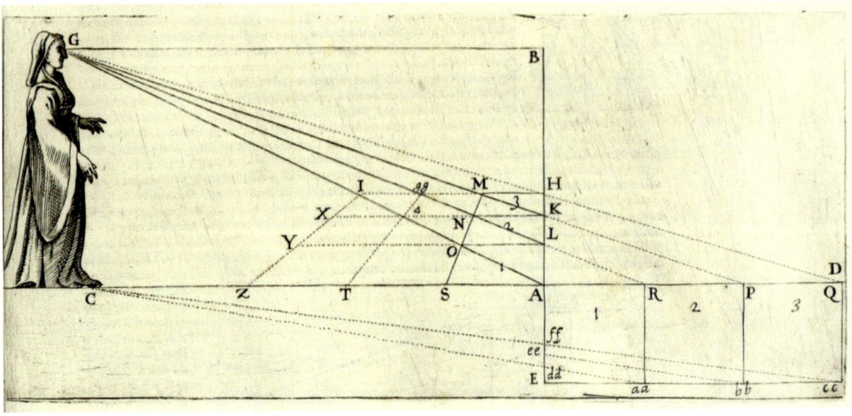

Figure 2.5 Illustration of perspective, Jacopo Barozzi da Vignola, 1583. Courtesy of Getty Research Institute, Los Angeles (93-B6464).

simultaneously touch the literal things of the world and to see past them into an abstract, axiomatic space. From its beginnings as a collection of approximate rules of thumb, perspective drawing assimilated an ideology of exactitude. Where Hooke's practice of exactitude was fidelity to the most intricate of details, Euclidean exactitude fostered by rational drawing methods excluded detail in favour of clarity. The eight-vertex brick was exact because it had been reduced to a calculable form. A precise line was defined by two points, with no variations in thickness or deviations from straightness. In drawings, thinness, fineness and smallness became aesthetic markers of precision.

Although the techniques of perspective have been largely settled since the late eighteenth century (and underpin contemporary drawing and modelling software), more recent understandings of light and vision have prompted new ways to imagine the space of sight. Leaping forward to the late twentieth century and its critique of drawing as representation, we find an altered version of the relationship between body and perspective, adjusted by modern relativistic physics. In his speculative project *Temple Island* (1987), Michael Webb, one of the founding members of Archigram, explored the extremities of perspective implicit in the idea of projecting lines to infinity, by introducing the relativistic effects of moving at the speed of light.[13] Webb imagined a viewer looking down the straight course of the Henley Royal Regatta, an annual rowing event on the River Thames. The vanishing point of this view would coincide with a neo-Classical folly designed by James Wyatt. The circular image, *11 O/P*, centres on this point (see Figure 2.6). On the left, the image incorporates a watercoloured landscape, tinted yellow (perhaps in reference to the paintings of Claude Lorraine (*c.* 1600–1682) and the celebrated Claude Glass).[14] The mirror-like surface of the water repeats the forms of trees, hills and the coloured wash of the sky. The

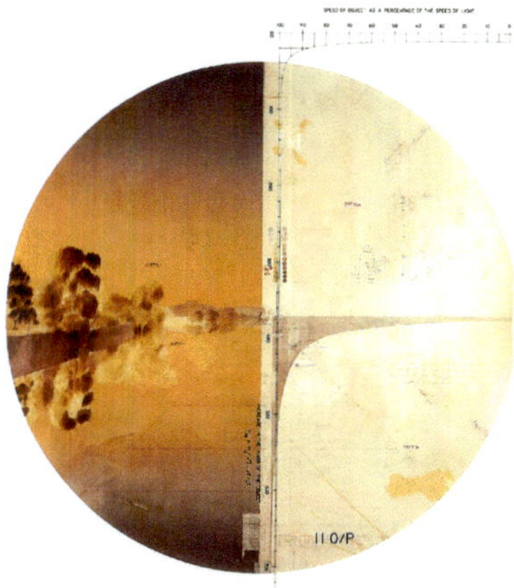

Figure 2.6 *Temple Island: Section through Cone of Vision, Circular Image*, Michael Webb. Image supplied by Archigram Archives 2024. © Archigram 1966–9.

right-hand side of the image is diagrammatic: converging lines and a circular graticule are used to graph the approach to infinity.

Webb intends for us to imagine the journey towards the vanishing point as an exponential increase in speed and a corresponding widening of space. Instead of a rowing skiff, Webb envisions the viewer, or 'tripper', reclining inside a sleek hydrofoil submersible – the 'Resurgam'. 'Webb, the Atomist', wrote Michael Sorkin, 'is looking for architecture's tiniest increment, the moment at which the whole complex field is contained in a single, primary dot'.[15] Like Dürer, Webb imagines perspective in physical, bodily terms: an elaborate arrangement of housings, postures and equipment that situates the drawer in the space of the drawing. Perspective drawing involves making or imagining full-scale three-dimensional spatial constructions which incorporate the bodies of both drawer and viewer. The three-dimensional arrangements of cords and lines used to construct early perspectives are not only a *means* to make drawings but are also themselves *already* drawings. The preoccupation with physically mapping objects and spaces in perspective requires exploiting the dual nature of lines. Even when these lines become intangible, as lines of sight, shadow or contour, the scene is converted into a graphic figure that bridges the infinite (the vanishing point) and the concrete (the point of contact). The perspective, even when constrained to a flat surface, implicates specific points in real space. The fundamental assumption of perspective is the *continuity* of space through the picture plane; a line can be drawn from the physical eye of the viewer into the notional geometric space of the picture. The viewer waits, suspended in the moment just before being precipitated into the image.

Scanned bodies

Human bodies proved particularly challenging, technically and conceptually, to set in perspective. Bodies were often drawn as if they could enter and interact with a geometric world, but were of a different order from it, modelled with shadow, materiality and posture (see Figure 2.7). Dürer sometimes depicts figures interacting with these lines: gesturing towards them, pointing through them or pinching them between thumb and forefinger. Renaissance painters commonly constructed an architectural frame using perspective but fell back on more traditional freehand pictorial methods to draw the human occupants of that frame. Despite technical and conceptual difficulties posed by setting human bodies in perspective, however, the growing desire for a rational, universal drawing system meant they too needed to be convertible into lines and points.

The pragmatic Dürer recommended approximating irregular objects first as a box which could be easily positioned in perspective, which would then guide the drawer to fill out details. He demonstrated how, by adjusting proportional guidelines within this frame, the drawer could produce various physiognomies and expressions.[16] Facial differences were not characterful exceptions to an aesthetic norm but configurations susceptible to a geometric synopsis. A commonly reproduced engraving of Dürer's shows another of his drawing devices, used specifically to capture a human body in perspective (see Figure 2.8). A framed picture plane holds a grid of wires or threads mounted halfway along a table. At one end of the table lies an awkwardly posed and partially draped woman; at the other, a grim, moustachioed artist squints at her, his viewpoint fixed by a

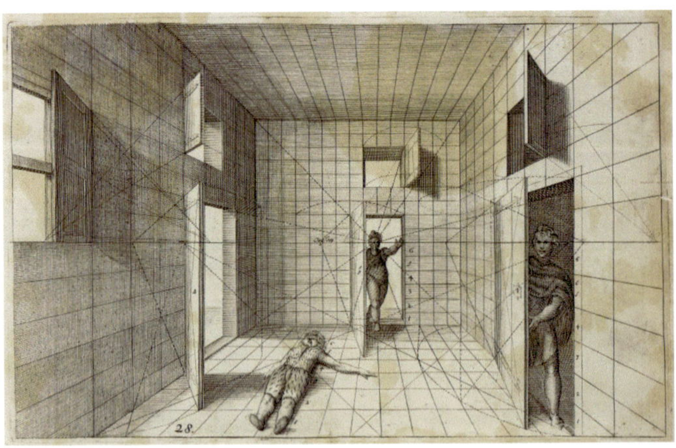

Figure 2.7 Perspective, part I: plate 28 (complex quadratura constructions incorporating human figures), 1604, Hans Vredeman de Vries. RIBA99467. Image courtesy of RIBA Collections.

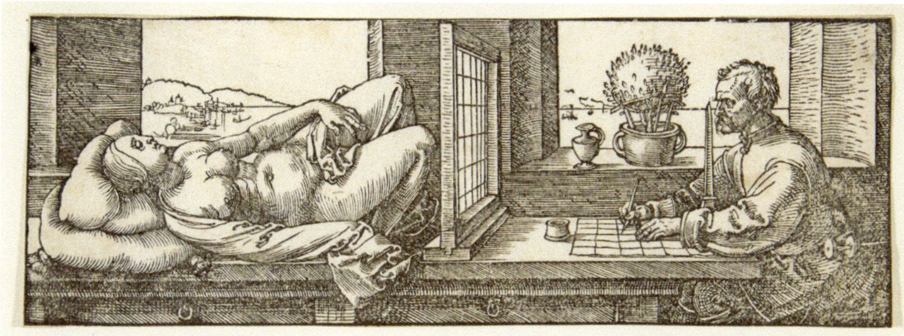

Figure 2.8 *Draughtsman Making a Perspective Drawing of a Reclining Woman*, c. 1600, Albrecht Dürer. Woodcut. Gift of Henry Walters, 1917. The Metropolitan Museum of Art Collection, New York.

miniature obelisk. Working one square at a time, he transfers his observations to a sheet of gridded paper. This setup relies less on pinpoint accuracy and more on the artist's ability to trace contours. The picture plane and the drawing are separated, with the artist as the link between the two. It is also worth noticing how strange the artist's view of the woman is: highly foreshortened, emphasizing her knees and thighs, the overall structure of her body de-emphasized in favour of her outermost contours. Perhaps Dürer intends to show us how this method is useful in situations in which the body is defamiliarized and one's knowledge of bodily proportion is less decisive.

Dürer's investigations of how bodies could be converted to geometry – through bounding boxes, proportional guides or contours – recall older techniques by Piero della Francesca (1415–92).[17] He, too, constructed his perspectives first as a physical, three-dimensional drawing-in-place, as Evans describes: 'Piero recommends stretching long filaments, such as hairs from the tail of a horse, from nails fixed in the eye positions, then marking intersections with the picture plane using numbers to correlate them. . . . This produces an array of dots, as did the cursors in Dürer's machine. Join up the dots and there is the image.'[18]

Complex forms, like the human body, could also be magically revealed in perspective by converting them into an array of dots. In a well-known drawing made prior to 1492, Piero drew a human head in plan and elevation as a series of sectional profiles defined by rings of points (see Figure 2.9). Converted to points, the head becomes a posable model to be rotated and positioned in space. Alberto Pérez-Gómez and Louise Pelletier compare these to a modern CAT scan, but one that exposed the geometric order hidden in living bodies.[19] Geometry was not so much applied to human bodies as revealed in them; bodies were not exceptions to the geometric order but its apotheosis. This 'act of conceptual surgery' rested on what we would now call *sampling*: using a small number of discontinuous datapoints to approximate a more complex continuous whole.[20]

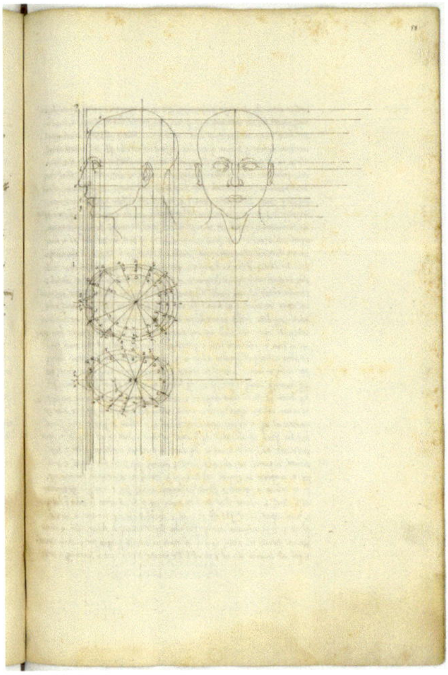

Figure 2.9 Elevations and sectional profiles of a head, Piero della Francesca, from *De Prospectiva Pingendi*, Petrus Pictor Burgensis. Image courtesy of Museo Galileo. Bibliothèque Nationale de France.

We do not know how Piero obtained his points. His contemporary, Alberti, devised an apparatus he called a *finitorum* to survey points from a body for sculpture or perspective; Evans speculates Piero used something similar.[21] The *finitorum* was a circular disk on an offset arm that could be positioned above the body to be drawn (see Figure 2.10). Plumblines were suspended from this disk until they just touched the body. Radial gradations on the disk and the length of the vertical plumbline allowed for measurements of the body to be read off in polar coordinates. To access a body's recesses, Alberti recommended using a rod (*exempeda*) as a probe. He breezily insisted his method made it trivial to measure a living human body and offered a table of measurements ('To the hole of the ear: 5 feet, 5 degrees, 0 minutes').[22] Painstaking labour by the artist and superhuman stillness on the part of the model must have been required to generate each precious point.

Dürer, and before him Piero and Alberti, scans the body, sampling it in various ways. The density or paucity of sampled points correspond to the practical labours involved in locating them. Living bodies are immobilized, probed and reduced to forms interpolated from a table of coordinates; this is the price of rendering them computable in Euclidean space. Exactitude requires holding things still, fixing them as a set of spatial coordinates. In each case, the mobile and flexible surface

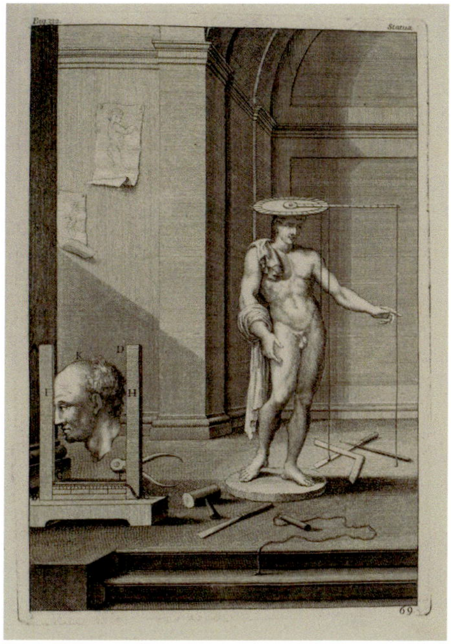

Figure 2.10 *Finitorum*, from *Della architettura, della pittura e della statua*, Bologna, 1782, Leon Battista Alberti, p. 332. Universitätsbibliothek Heidelberg. https://doi.org/10.11588/diglit.1558.

of the body is figured as a network of fixed points, each of which can be located precisely in relation to one another. The more finely spaced the points, the more detail appears. The obelisk, eyepiece, eyelet and plumbline all eliminate movement; with the model held stationary, the points can be measured as if simultaneous. The model posed in the *finitorum* reminds us of artist Rebecca Horn's (1944–2024) *Measure Box* (1970) which makes explicit the intrusiveness and power dynamics involved in such conversions (see Figure 2.11). Where perspective, however, defines form in relation to viewpoint, the scan translates its object into form without viewpoint.[23] The geometrically transfigured body was given a new kind of mobility; it could be rotated, transformed, posed and positioned.

The method of scanning objectifies things by desemanticizing them and subjecting them to a coordinate system. It reads surfaces from close up, unconscious of the meaning or form of the whole: 'Scanning passes over things evenly, distractedly, objectively, it is as tactile as it is visual, its mediation is a point, not a line, and the gaps consist of what lies outside thresholds of detection as much as what hides behind visible surfaces.'[24] Modern lidar scanners are fundamentally a species of *finitorum* that re-enact, automate and accelerate Alberti's method using a laser to measure distance: the time taken for emitted light to reach a surface and bounce back is measured and combined with the

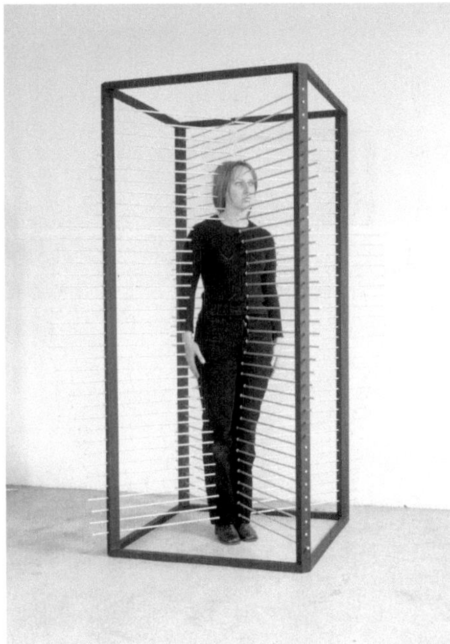

Figure 2.11 *Measure Box*, 1970, Rebecca Horn. © Rebecca Horn. VG Bild-Kunst/ Copyright Agency, 2024.

known three-dimensional angle at which the laser was projected in order to arrive at a position or point. If a laser is as close to an abstract geometric line as it is possible to physically create, the resulting point is formed by a Euclidean intersection between point and surface. The head of the lidar scanner rotates at high speed taking millions of such measurements – it exploits and trusts its ability to blindly sample a large number of points to provide the required information – resulting in a table of coordinates, measured as polar coordinates (an angle and a distance from a known point), able to be converted into Cartesian coordinates.[25]

When these are plotted three-dimensionally they produce what is known as a point cloud, a set of data points in a three-dimensional coordinate system. A common visual effect for point cloud imagery is to present the data as bright points against a dark background, as if it were a night sky (see Figure 2.12). Lidar scanning can be understood as a kind of spatial tracing which results in images with ephemeral, veil-like qualities. A video installation, *Te Ahua, Te Wa, Te Atea* (2022), by Aotearoa New Zealand artist Russ Flatt (Ngāti Kahungunu), uses this effect to evoke the possibility of non-human vision: we are taken over areas of land 'as if we are scanning like the kāhu or hawk'[26] (see Figure 2.13). The swellings and depressions of the ground appear as a weightless constellation of

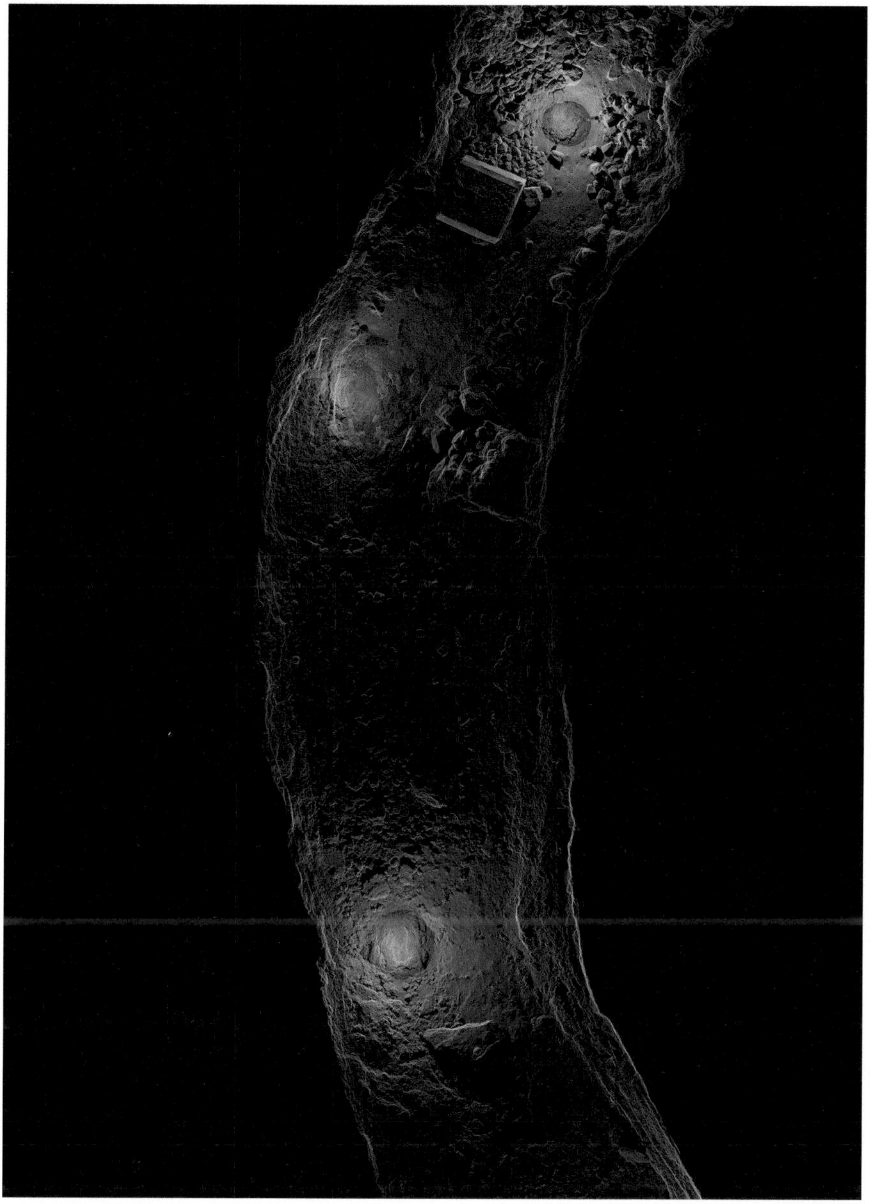

Figure 2.12 *Mount Royal Cave*, from the series 'Nga Mahi Rarowhenua', 2018, Chirag Jindal. © Chirag Jindal, Terrestrial LiDAR Scan (2018).

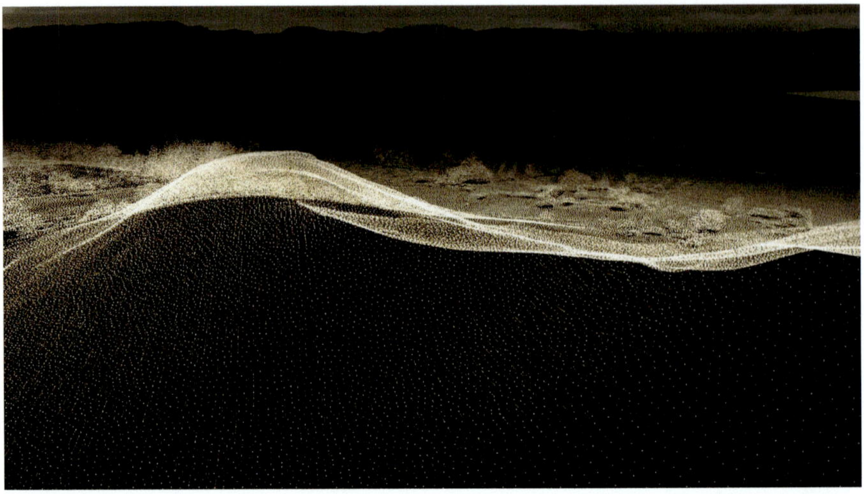

Figure 2.13 Still taken from *Te Ahua, Te Wa, Te Atea* (8-minute single channel video installation work), 2021, Russ Flatt. Russ Flatt c/o Tim Melville Gallery, Tāmaki Makaurau Auckland.

points, moving as the virtual camera pans and rotates. The lidar scanner, a blind surveyor, moves without respect to this encoding, superimposing its own radial order. Flatt exploits this alienating effect to evoke the sight of a bird, an ancestral presence or the witness of the land itself. The kāhu, like a drone, hovers over the landscape in a search pattern, plummeting to impale its prey against the ground or alighting on already-immobilized carrion.

Lidar scans are particulate. Michael Young observes that point-scanning techniques cannot capture edges or lines but only clouds of points that approximate them: 'Edge in this context is thus not the geometric intersection traditionally represented and controlled through the graphic drawing of a line. Edges in three-dimensional scans often look odd; they smudge and fray, sometimes drifting into the shadowy void.'[27] The points of a scan result from touching. A sharpened probe, be it a plumb-bob, stylus or laser, is extended to make contact with a surface to yield measurements that fix the point in space. Scanning makes the geometric world touch the physical one, allowing geometric figures to be treated as the abstract, essence or experimental analogue of physical things.

Sampled terrains

Instruments and techniques for scanning bodies closely resemble methods for surveying terrains. In each case, the problem is how to render complex forms geometrically. Once again, the solution involves sampling forms as a collection of fixed points, each of which represents a scaleless touchpoint between a material world and a geometric one. By touching the former precisely enough, it could be converted to (or revealed to already be) the latter. Edmund Husserl went so far as to claim that all geometry originates in surveying.[28]

To survey Rome in the 1440s, Alberti employed a *goniometer*, a device very similar to the *finitorum*. On an elevated point that he specified as the city's centre, he positioned a level drawing table inscribed with a 10-foot-diameter circle – the *horizon* – establishing a coincident centre for drawing and city. Sighting along an alidade – the *radius* – Alberti took bearings of significant locations, circling his plane table as he drew distant features of the city onto it. Significantly, Alberti published his map not as a drawing but as raw data: a table of coordinates. Each user of the data, he argued, could produce a map suited to their own needs (much as he did for his *finitorum*). It is not clear how he measured the distances to each survey point; historian Jessica Maier suggests he simply paced them out.[29] Additionally, some of the points he located would not have been visible from his central point, so it seems likely he consolidated data from several survey locations, transforming it to appear as if taken from the Capitoline Hill (the seat of government, if not the geographical centre of the city).[30] The data is full of gaps; relatively few points are measured. This may well be deliberate however: by reducing the city to a brief list of strategic points, he offers an administrative view, rendering the city a project. Maier calls his drawing a 'geometric framework . . . a flexible platform that others could expand and embellish as the modern city was transformed'.[31] A synopsis, however, was only achieved by walking and stretching lines across the actual city. Alberti generated a geometric city by means of an act of full-scale drawing that isolated and connected key points.

Such drawing acts could reach further, apprehending the globe itself. In *On the Art of Building*, Alberti claims the curvature of the earth can be surveyed ('ten inches in every mile') by setting up posts of known length on flat ground and sighting along them.[32] Alberti writes:

> Those who have studied the matter declare that the earth is spherical, although much of it is wrinkled by mountains and much of it clothed by sea; yet with a globe so large the roughness is scarcely noticed: it is like an egg that, although not smooth, has projections that seem minute and insignificant when compared to the size of its circumference.[33]

Although it appears irregular, the earth becomes geometric at very large and small scales: the globe and the point. The surveyor, drawing lines across the earth, connects these rational worlds. The tension between rough ground surfaces and their geometric abstraction recalls Hooke's scalar imagination: distance provides the abstraction necessary to see the earth as a sphere, or a blot as a point. It triggers imaginative shifts of scale, so that both the infinitesimal and vast can be imagined as inhabitable. Hooke was himself a notable surveyor. After the Great Fire of 1666 reduced much of the city to a charred heap on the Thames, he was appointed Chief Surveyor (reporting to his close collaborator Christopher Wren (1632–1723)) in charge of mapping and laying the ground for rebuilding it.[34] The work exercised his skills of measurement and tabulation but also called for considerable diplomacy as he negotiated with landowners to agree on boundaries and enact the government's requirement to widen existing roads. It was also physically demanding work. Once landowners had cleared the ground and revealed the foundations of their former buildings, Hooke and his team of surveyors examined the sites to confirm their boundaries, hammering deep stakes into the ground. These stakes, protected under law to discourage tampering, were the pins anchoring the living territory to an emerging geometric diagram. Between these fixed points Hooke stretched cords, tangibly manifesting the boundary lines that confirmed properties' statutory existence. These slack strings were pulled tight until straight and so linearity emerged through tension. These lines were then measured, recorded, dated, located and sketched into the cadastral record.

Hooke's survey was effectively a full-scale drawing project to prick out the plan of London over its burned foundations. Surveying is commonly understood as a *means* to make a drawing. Recorded information can be used to produce two-dimensional inscriptions, scaled for various uses. But the survey could also be understood as *already* a drawing, spatially enacted by the surveyor, transient and incremental. Such drawings may be compared with the constructions and diagrams used to effect the transparent window of perspective. They also echo the use of stretched rope-lines used to set out construction sites. Although transient, the lines of Hooke's survey drawing are not gone; they persist, materialized as built fence-lines, building edges, kerbs and walls that remain in place today. The survey drawing opens space; a line 'transforms the material surface of the paper into a virtual space, an open background' as much as it creates foreground figures.[35] The point, the first contact, is not only the setting-out of the drawing but also the establishment of the field that will remain empty. In the surveyor's drawing, the spaces between the network of lines and points are prepared for action, made saleable, taxable and defining the space of property rights within a bureaucratic system. Surveying empties out space by way of drawing, preparing it for use, occupation or appropriation.[36]

While London was being rebuilt on Hooke's survey plan, he also designed and built a monumental geometric instrument. The Monument to the Great Fire of London (1671–7), a fluted column on an elevated base topped with a bronze urn, still stands at the junction of Monument Street and Fish Street Hill near the River Thames. It marks space like a post or survey peg and offers an overview of the city.[37] The vertical height of the monument was determined by its horizontal distance from the bakery where the fire began, making it the vertical side of a right-angled isosceles triangle. The tower was also conceived as a geometric apparatus. Set into the column's foundation was a small laboratory space and within its centre, a spiral staircase with internal void; a hatch inside the urn could be opened to reveal the sky. With lenses mounted in this shaft, it could be used as a giant vertical zenith telescope for observing stellar parallax (see Figure 2.14). Hooke intended that from a supine position in the basement, his eye could intercept a line between the earth's centre and the distant stars. Minuscule variations would reveal the mysteries of the earth's orbit. There is a kind of symmetry in Hooke's work: a viewer peering through a microscope at the world revealed in a full stop was mirrored by the viewer gazing up at the pinprick stars.[38]

Points – whether marked by sharpened stakes driven into the ground, a pin in the wall with attached cords or a stylus brought into contact with a model's skin – focus thoughts and aspire to precise correspondence between real and virtual domains. A drawing apparatus or setup – for producing perspective, scanning bodies and surveying terrains – is not only a means to an end but also a space the drawer inhabits. In these spatial constructions, drawing instruments, drawing acts and drawing marks are indistinct. Bodies were understood to be *within* and

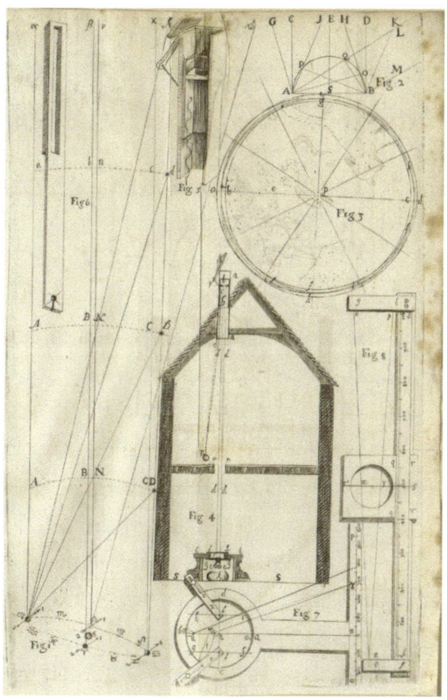

Figure 2.14 Drawing of an azimuthal telescope, 1674, from *An Attempt to Prove the Motion of the Earth from Observations Made*, Robert Hooke. Library of Congress, Rare Book Special Collections Division.

situated by geometric constructions. Being sizeless, a point in the material world is equally at home in the context of the infinite and infinitesimal.

Touchscreens

Perspective and surveying are geometric techniques and drawing practices that allow exchange and conversion between a material world and an abstract geometric one. In the various spatial constructions considered here, we notice how the links between drawing instruments, drawing acts and drawing marks are blurred. Dürer's physical perspective apparatuses and the web of lines constructed by Alberti's *finitorum* are – in our view – not just tools for drawing but already drawings in themselves. In this we admittedly project back an understanding of drawing that does not emerge until the mid-twentieth century (as discussed in the previous chapter), but it is not entirely anachronistic. As Paul Emmons points out, many key features of architectural drawing have their roots in the construction site or are in a close imaginative dialogue with them.[39] Architectural drawing is intrinsically a situated, enacted, bodily practice that only achieves abstraction and autonomy through considerable labour. Geometric techniques are at the root of the rationalization of drawing and subsequently its digitization. The idea of a computational analogue – a digital model – builds on the way design drawings were understood to define and constrain their future realization. Design software explicitly manifests a Cartesian understanding of space and generates orthographic and perspective views according to the tenets of descriptive geometry. If anything, computers only sharpen the disjunction between Euclidean and Hookean worlds which comes to be seen as a matter of *interfacing*.

The now-ubiquitous touchscreen, for example, filters the smears of blunt digits, converting them into a skeleton of data which can be rendered on the screen as if the finger were a fine ink pen, a watercolour brush or a carving tool. The velocity and pressure of touch can also be quantified and used to vary effects. The touchscreen renders us digital so we can act in digital space. Algorithmically sharpened, the finger becomes a precise instrument. And the screen touches back. A surprising variety of tactile experiences can be produced by well-positioned oscillators behind the screen, mimicking the feeling of clicking a button or running a finger over a textured surface. Touchscreens orchestrate the sense that we are interacting through a thin interface separating us from the world of data. Far from a new invention, touch-based interaction with computers is at the origin of computer drawing. The earliest system to use a complete graphical user interface, Sketchpad, developed by Ivan Sutherland in 1963, used a light pen to locate points on a screen as if it were a page to draw

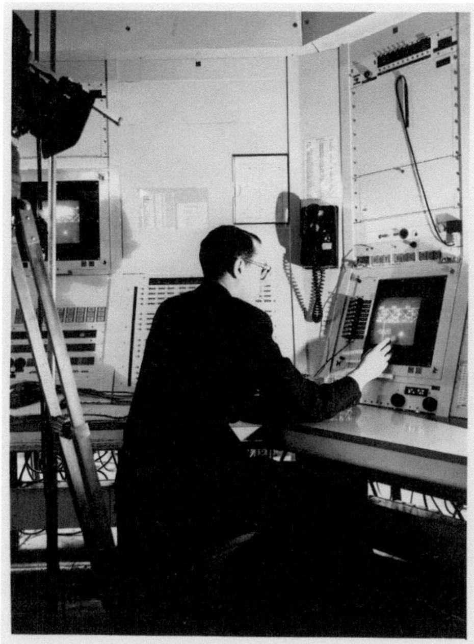

Figure 2.15 Demonstration of Sketchpad, 1963, Ivan Sutherland. Image courtesy of MIT Museum.

on[40] (see Figure 2.15). As the operator moved the pen across the glass display unit, a cross-shaped mark appeared, designating the current operating location. Pen tip and cursor were kept synchronized by the refresh rate of the screen. The system gave the effect of information passing back and forth through the screen, as if it were a mere membrane between corporeal and virtual space. The cursor was programmed to have what Sutherland referred to as a 'gravity field' that would enable it to snap to nearby locations of significance, such as the end of a line or the centre of a circle.[41] The operator could work approximately and rely on the computer to maintain precision. Sutherland described the interface as a way 'to converse rapidly through the medium of line drawings' with the computer.[42]

The metaphor of the screen as a window perpetuates Alberti's understanding of perspective. Drawing is not so much a set of marks that correspond with an external reality as it is a screen or portal set up between viewer and viewed. The drawing is intercepted at the boundary between two spaces. The act of drawing passes through the surface rather than across it. The sense that we can see through the page-as-window, that the computer is a conversation partner and that we can use our fingers to manipulate pure geometry are effects of the interface and the way it presents itself to our eyes and hands.

Through the window

Euclid's and Hooke's points have stood in here for the distinction between rational and embodied modes of drawing. Employing these as representative, we emphasize the ways in which architectural drawing is fundamentally a knowing negotiation, a constructive dialogue or an interface between these two modes. Architecture's reliance on drawing practices depends on being able to bring these abstract and material worlds into contact and operating simultaneously in rational and embodied modes. *Transparency* is an effect of this interface. It is the sense that we can move smoothly and directly between two realms. Various conceptual and practical ways have been developed to understand the directness of this movement, such as representation, windows, screens, metaphors, translation and projection. To enable transparency, something must pass through the window, and the spaces of body and image must connect in a predictable way. Transparency is a visual metaphor, but it is often associated with tactility; the most transparent interaction is physically direct. To be precise might be referred to as putting your finger on it, and the labours of working precisely often rely on sharpening some instrument as finely as possible. As Young argues, the persuasive quality and practical efficacy of a lidar scan is the sense that it has connected with the surfaces it records: 'The information captured, stored, and processed through lidar and photogrammetry is not only trying hard to see as many qualities of environmental surfaces as possible, it is also trying to give them spatial depth, in a way, it is trying to *touch* reality.'[43]

Points are crucial in the synoptic handling of form. By reducing the complexities of real material objects to a relatively small number of points, they become manipulable in the rational logic of constructed drawing. We might now recognize Piero's method of measuring heads as a kind of data compression and see the close similarity between this and Alberti's process of surveying Rome. Parsimony was essential: a balance was needed between specifying enough points to capture form in appropriate detail and specifying few enough that they remain graspable. Drawings that became overwhelmed by points could become lost and opaque. Space opens between points; geometric synopsis empties out space. In reference to the reciprocity, and hence essence, of drawing as being between existence and inexistence, Alain Badiou writes,

> In one sense, the paper exists, as a material support, as a closed totality; and the marks, or the lines, do not exist by themselves; they have to compose something inside the paper. But in another and more crucial sense, the paper as background does not exist, because it is created as such, as an open surface, by the marks.[44]

Similarly, artist Richard Tuttle wrote, 'The paper is just as much drawn as mark is drawn.'[45] This is more than a visual effect: the emptiness of the space between points is transferred back to the territory so that it also appears empty, available for use, waiting to be filled.

In this chapter, we have considered how the transparency of the drawing (i.e. its ability to put two places in contact with one another) is constructed first via full-scale acts of drawing. We employ the notion of drawing as a performative, material enactment – as a verb – from the critical expansion of drawing by artists in the second half of the twentieth century. Reflecting on past scenes crucial to the development of rational (and retrospectively pre-digital) drawing, we interpret Dürer's perspective apparatuses and setups, and the physical work of surveying, as full-scale drawings. In doing so, the distinction between instruments of drawing, acts of drawing and artefacts of drawing blur. Drawing precisely required positioning the drawer's own body by means of points in space. As Emmons puts it, 'the drawing table is the pivot between the physical body of the drafter and the imaginal body projected into the design'.[46] The full-scale drawings in this chapter are investigations or inquiries, attempts to disclose, discover or work something out.

The 1:1 scale drawing has frequently been a satirical device. In Lewis Carroll's *Sylvia and Bruno Concluded*, a character asserts the existence of a full-size map but admits that it cannot be unfolded: 'the farmers objected: they said it would cover the whole country, and shut out the sunlight! So we now use the country itself, as its own map, and I assure you it does nearly as well'.[47] A similar map is made in Jorge Luis Borges's (1899–1986) brief parable, *On Exactitude in Science*, with similar results: 'The following Generations, who were not so fond of the Study of Cartography as their Forebears had been, saw that the vast Map was Useless, and not without some Pitilessness was it, that they delivered it up to the Inclemencies of Sun and Winters.'[48] At the risk of over-explaining a joke, the full-scale map is shown to be impractical because it offers no synopsis and is too arid for inhabitation. A scaled drawing differs not only in size but also gains its value from its level of portability and abstraction. In a Euclidean world, scale is a mathematical transformation, a multiplication. For Hooke, scale is a magnification that enables us to draw close. As we draw closer, however, we paradoxically move further away from ideal geometry. Through his downward-facing microscopes, and the telescopes he built to look upwards, Hooke imagines something we would now recognize as 'zooming', the smooth increase or decrease of magnification to produce the effect of changing distance. Webb dramatizes the scaling implicit in drawing: his scaling is accelerating motion tending to infinity, the sort of movement that requires us to be nestled in a life-support system; it is worth remembering that the term *zoom* originates in the onomatopoeic memory of a rushing sound.[49] The expanded approaches to drawing in the

1970s, discussed in Chapter 1, did not simply introduce something new to architecture. They reactivated much older possibilities of drawing, precisely at a moment when those possibilities were also being given novel form by emerging digital technologies. Something similar can be observed in the relationship between drawing, gesture and movement, to which we turn in the following chapter.

3
DRAWING IN TIME
TRACKING LINES IN MOTION

Performed lines

A drawing is a record of something that happened; to consider drawing as a situated, bodily activity is to pay attention to how it unfolds or feels its way ahead. Drawing-in-place is also drawing-in-time. Architectural drawing, in its usual context as a prescription for construction, fixes things in space and time, determining the locations of a building's elements and in what order and over what period of time they appear. Such drawings prescribe a complex choreography of human bodies, machines and materials. To do this, however, they suppress the time of the drawing itself. Omitted also is the enactment of the building process that the drawings set out, just as the surveyor's journeys back and forth across a terrain are excluded from the geometry of the survey. In the previous chapter we reflected on drawings that are body-scaled instruments for fixing things in place, locking the relationship between drawing and thing-drawn. In this chapter, we turn our attention to drawings that are corporeal and expressive: bodies performing lines.

Since the widespread arrival of computers within the practice of architecture in the 1990s, the place of the drawer's body has often been deprioritized in favour of standardization, algorithms, immateriality and production. While software designs and conventional uses tend to reinforce this, it is however, avoidable. We suggest that computational architectural drawing can also be seen in the context of expanding drawing practices, yielding new possibilities for drawing rather than making it obsolete. The understanding, which occurred in the 1970s, that drawing is an embodied, situated and performative act met with an intensifying interest in valuing architectural drawings independent of their role in producing buildings. Out of this sprang a new critical understanding of architectural drawing, characterized by speculative drawn projects, hybridized drawings and models and an increased importance of exhibition and theoretical analyses of representation. Two new genres of art that further developed in the 1970s proved influential: installation art – discussed in Chapter 1 – and performance art, considered here. In terms of drawing, Cornelia Butler and Catherine de Zegher describe this period as one in which 'Line was taken for a walk into the "real" world – onto the floor, the walls of buildings, into the plains and mountains, close to the earth, high in the sky.'[1] To begin, let us consider several works of performative drawing from this time.

In 1974, Aotearoa New Zealand artist Jim Allen (1922–2023) presented a three-part performance at Auckland City Art Gallery, culminating in *Contact: Body Articulation/Imprint* (see Figure 3.1). As Allen recalled over a quarter century later:

> We covered the walls with plastic. The performers were instructed to start off by exploring their own body articulation by rubbing on paint and flexing, and then to extend their bodies in different directions. Most of them did that by lying on the floor and circumscribing movements with their arms and legs with paint. They spent a bit of time doing that as individuals, and then formed pairs to explore the physical relationships between two people, extending all their movements. People were picked up by the ankles and dragged around in big circles in paint. As *Body Articulation* progressed, the audience became far too enthusiastic and I was worried they were going to get their clothes off and get in amongst it. Seriously. And I thought if they did that, the Gallery would close it down. There would be a riot and that would be the end of it. It almost got out of control.[2]

Allen's glee at this unruly scene is barely suppressed. The floor and walls of the room became a three-dimensional canvas and the bodies of the participants a wild drawing apparatus. Smeared marks directly indexed the friction and weight of flesh, the arc of an arm's swing, the texture and creases of skin. This performance resulted in the print of a composite, collective body in motion which remained on show for the duration of the exhibition.[3]

At around the same time, Carolee Schneemann (1939–2019) suspended herself in an arborist's harness. Hanging and twisting, she drew on the walls and floor in crayon, mapping the constrained movements of her body to produce *Up to and Including Her Limits* (1973–6) (see Figure 3.2). Schneemann explains that she became an extension of the rope itself and, trance-like, 'let my entire body function as a pencil'.[4] She did not so much wield a drawing implement as become one. Drawing traditions have often emphasized skill and nuanced

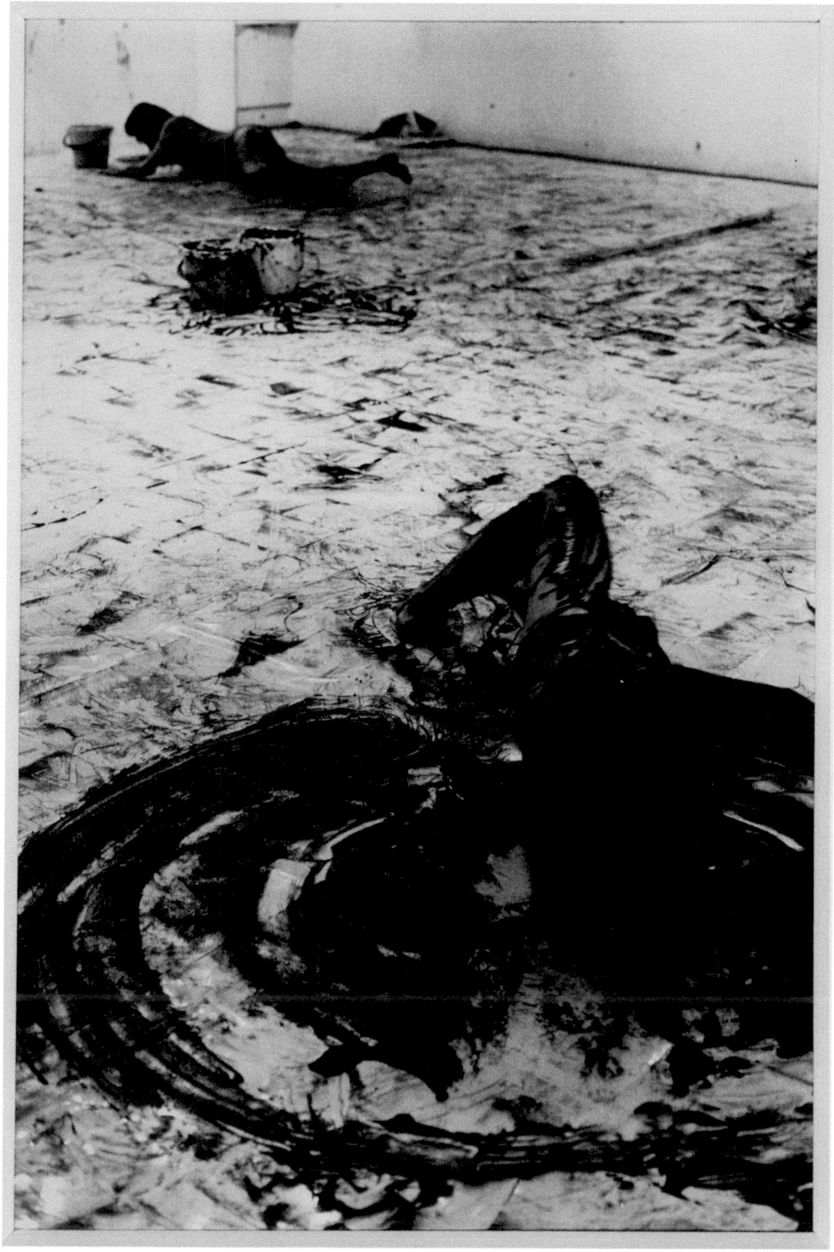

Figure 3.1 *Contact: Body Articulation/Imprint 6*, 1974, Jim Allen. Silver gelatin print. 9 3/8 in × 6 19/64 in (238 × 160 mm). Photograph by Bryony Dalefield. Auckland Art Gallery Toi o Tāmaki, purchased 2013.

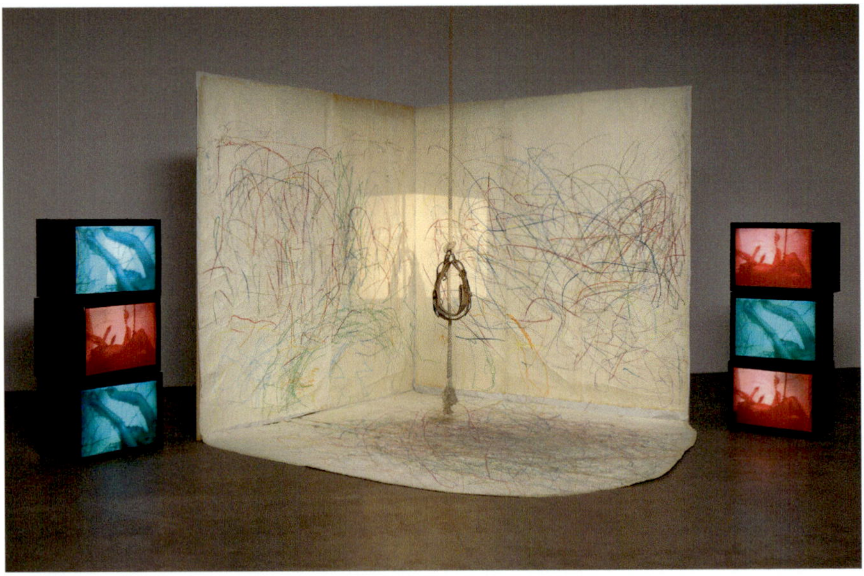

Figure 3.2 *Up to and Including Her Limits*, 1973–6, Carolee Schneemann. Crayon on paper, rope, harness, 16 mm film projector, video (colour, sound; 29 min.) and six monitors. Dimensions variable. Committee on Drawings Funds and Committee on Media and Performance Art Funds. Acc. no.: 520.2012.a-j. Digital image, The Museum of Modern Art, New York/Scala, Florence. © Carolee Schneemann. ARS/Copyright Agency, 2024.

control but for Schneemann the human body's limits and fleshy scope make it a clumsy, aleatory practice.

In *Time Line* (1968), Dennis Oppenheim (1938–2011) physically manifested an invisible borderline. Part of the border between Canada and the United States follows the centreline of the St John River in Maine. The border also marks the meeting place of two time zones; crossing such statutory and temporal lines requires re-synchronizing oneself to a new jurisdiction. In winter, when the river was frozen, Oppenheim walked along this border and dug a mile-long, 3-foot-wide line in the snow and ice.[5] This line is site-specific, drawing out something concealed in the landscape.

Choreographer Trisha Brown's (1936–2017) *Man Walking Down the Side of a Building* (1970) sets a performer in an abseiling harness to descend the face of a brick building from the roof to the street, painstakingly constraining his movements as if he were walking on the ground (see Figure 3.3). The act of walking gives postural and gestural clues to the presence of gravity, balance and momentum. To walk down the face of the building required immense control to maintain these clues: 'the effort of looking effortless'.[6] While Oppenheim's *Time*

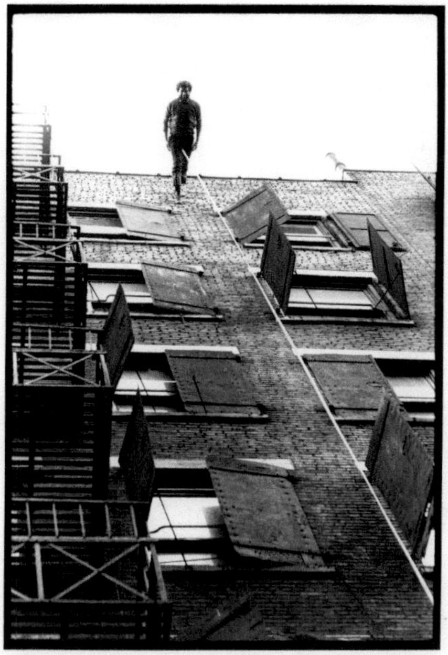

Figure 3.3 *Man Walking Down the Side of a Building*, 1970, Trisha Brown. Photograph by Peter Moore; © Northwestern University. Courtesy of Peter Moore Photography Archive, Charles Deering McCormick Library of Special Collections, Northwestern University Libraries.

Line (and other famous walking lines of the period, such as Richard Long's *A Line Made by Walking* in 1967) attended to the visible trace of the line, Brown's exists only in its fleeting performance and the filmic record. It is drawn yet disappears each time the work is performed.

These four works are aligned with performance art and its ideas of enactment and theatre; their focus is immediacy and how each piece is created. In Chapter 1, we observed examples of drawings leaving the page or wall, culminating in vigorous practices of installation-drawing emphasizing the tangible, material qualities of lines. Performance art, another major artistic paradigm to develop concurrently, became entangled with this conjunction of drawing and installation. It drew attention to the enactment of art, to art as something that occurs – a process or happening – rather than existing. In considering these four examples beyond their theatricality, for which an audience is crucial, drawing emerges as performative: focused on bodily movement, lines are created by bodies' traces, tracks or records. Allen's writhing bodies leave behind an eloquent mess; Brown's dancer suppresses the visual signs of effort while Oppenheim expresses them; and Schneemann awkwardly stabs at the walls with 'anti-gravitational delight'.[7] They interrogate architectural space by posing the body as unruly, irregular, difficult to control, critical, unpredictable and, most importantly, moving. By centralizing the body, acts of drawing-in-time explore flux and duration rather than fixity.

The expansion of drawing in art and a desire to valorize architectural drawings as separate from building production raised crucial questions about architectural drawing's representational paradigm of transparency and projection. Performative drawing does not refer to, or project, other places and times. While they may be performed again, they are not identically reproducible, since they express the result of a particular temporal situation. We explore the idea of fleeting, fugitive drawing practices within architectural contexts so that drawers' bodies are highlighted and duration exposed.

Sketch as captured gesture

Gestural sketches have long been taken as innately free expressions: an individual, personal and emotive mode of drawing. In the nineteenth century, Robin Evans (1944–93) tells us, it was assumed that 'the unpremeditated and unregulated' were 'signs of art and feeling'.[8] In contrast to the technical procedures of orthographic drawing and perspective construction, freehand drawing was understood and taught as a matter of habit, intuition and tacit skill. Gestural freehand drawing became an important element of design education. At the start of the twentieth century, the Debschitz School in Munich opened; here, drawing was not copying or imitating, but seen as 'an embodied re-experiencing' which students *felt* as much as saw.[9] In his pedagogical notebook, the school's co-founder Hermann Obrist (1862–1927) catalogued these qualitative lines: 'the looming, the lying, the receding, the striving, supporting, plummeting, undulating, tilting, falling, subsiding, soaring, bending, winding, resting, broken, cracked, gazing, defying, looping, striving toward, striving away, forward, backward, swirl, spiral attained by means of the intensification of the black and white, of massing, of the force of line, and so on'.[10] Paul Klee (1879–1940) (who worked briefly as an assistant at the school) took the gestural line further, famously intuiting that a line was a point in motion.[11] Lines projected from point to point were, he believed, 'more like a series of appointments' than a gesture or journey.[12] Any sense of travelling, exploration or response was lost once the line became merely the shortest distance between two points. The sketch also expressed things unintentionally. The gestural line was understood to disclose the unconscious, accessing emotion, spirit or collective memory. In contrast to Euclidean lines that jump between points, gestural lines carried nuanced expressive possibilities, particularly in terms of their temporality. The gestural line indexes time or can be read as if it does.

Insofar as the sketch is specific to an individual, it could also act as a kind of signature. The expressive sketch that condenses and foreshadows a completed building, as if it sprang fully formed from the architect's hand, is a well-worn cliché; to produce a building requires progressively more definitive and unambiguous sets of drawing, usually made by a team of people. The sketch, however, creates the illusion that an entire building might translate a single gesture, the work of an individual hand. Le Corbusier's (1887–1965) plump, sweeping lines, for example, moved easily between his mural paintings and his buildings, often translated as outlines of form in plan or section. He used a blackboard wall in his studio for large freehand sketches or making full-scale details to facilitate design transition. Sooner or later, however, the sketch leaves the notebook, blackboard or napkin. Through the development process, the fluid, conceptually intense sketch ramifies into a fully-fledged project – or so the narrative goes: the gesturally performed sketch is locked in place as a template to be traced, adapted and crystallized into building.

A sketch *captures* motion. Movement, particularly as revealed by film, was important to the imagining of modernity and modern architecture. New spatial types at the end of the nineteenth century – train stations, department stores, arcades and exhibition halls – were notably transient, linked by mobility. Film offered a new way of seeing adequate to this endless activity.[13] The pre-cinema studies of Eadweard Muybridge (1830–1904) and Étienne-Jules Marey (1830–1904) showed how the camera could reveal hidden intricacies of gesture and movement. Although their methods were different, they both adopted from orthographic drawing the principle that points could be located in space using paired views. Photographs taken from two directions could describe movement in three dimensions. Trajectories drawn through space could be arrested and examined like lines drawn on a surface (see Figure 3.4).

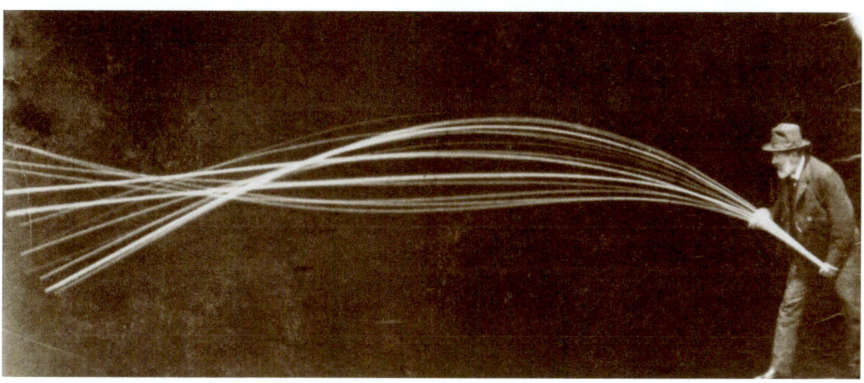

Figure 3.4 *Vibration of A Flexible Rod*, 1886, Étienne-Jules Marey. New York Public Library/Science Photo Library.

This technique for freezing motion quickly found a use in analysing and optimizing the repetitive movements of workers. Lillian and Frank Gilbreth (1878–1972; 1868–1924), industrial engineers and efficiency consultants, made pioneering photographic time and motion studies of manufacturing and clerical workers, beginning around 1910.[14] Their 'stereochronocyclegraph' technique used small electric lights on a subject's wrists that blinked twenty times a second; these registered on film as a punctuated linear trace[15] (see Figure 3.5). The Gilbreths believed that a worker's psychology and skill could be concluded from the structure of these lines; for example, undulating lines represented physical, and likely mental, hesitation due to inexperience or lack of habit.[16] Some of these lines were converted into wire models: three-dimensional drawings that abstracted and concretized fleeting motions (see Figures 3.6 and 3.7). The aim was to identify and eliminate inefficiencies in repetitive processes for ease, productivity and profit. Gestural lines revealed bodily variations for the sake of suppressing irregularity. In their photographs, a worker's body appears as a vague smudge and their movement like handwriting in need of improvement. This analytic-corrective treatment of the body soon seeped out of the factory and into the domestic realm.[17] In *The Home-Maker and Her Job*, Lillian Gilbreth offered housewives a means to study their own movements in the absence of a chronocyclegraph; Laurel D. Graham describes this study: 'This involved

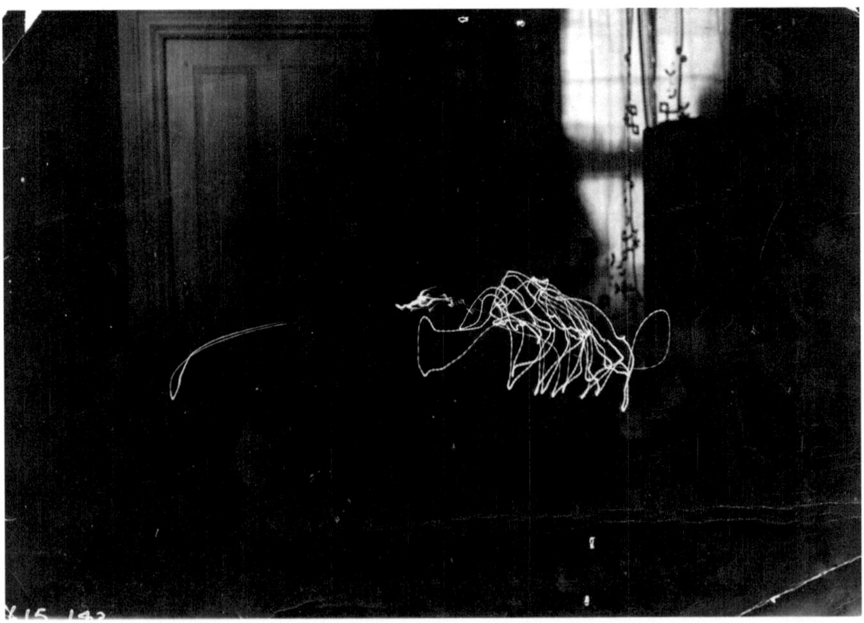

Figure 3.5 Chronocyclegraph of surgeon sewing, 1915 estimated, Frank B. Gilbreth Motion Study Photographs. Image courtesy of Kheel Center, Catherwood Library, Cornell University Library.

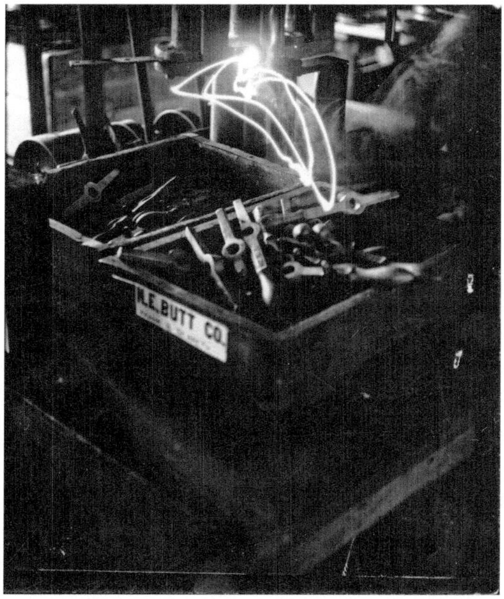

Figure 3.6 Two cycles on drillpress, 1915 estimated, Frank B. Gilbreth Motion Study Photographs. Image courtesy of Kheel Center, Catherwood Library, Cornell University Library.

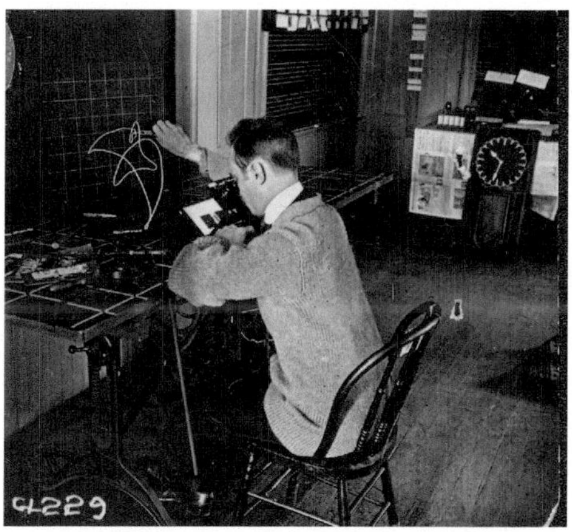

Figure 3.7 Making wire models from stereocyclegraphs, 1915 estimated, Frank B. Gilbreth Motion Study Photographs. Image courtesy of Kheel Center, Catherwood Library, Cornell University Library.

recruiting a child to follow the homemaker around the kitchen carrying a ball of string and leaving behind a trail to record the distance walked and a series of pins marking the points in space where she had turned.'[18] By constructing her own drawing in space, the homemaker would reveal latent movement patterns, disclose unnecessary actions and identify opportunities to ease her labours. While a Corbusian architecture of movement emphasized the visual unfolding of space around a moving occupant – the architectural promenade at the heart of the Villa Savoye is a stylized gesture retraced by the moving occupant – Lillian Gilbreth saw a looser fit between the interior and the moving occupant. What mattered was not the single line but the averaging of multiple lines of movement.

Apart from their practical uses, photographic motion capture had evident expressive possibilities. In 1949, *Life* magazine sent photographer Gjon Mili to visit Pablo Picasso (1881–1971) in southern France. There, Mili showed the painter a photograph he had taken of an ice skater, with tiny lights attached to her skates, in the manner of the Gilbreths' studies. These traces of light recorded her moving on the ice in the dark. Picasso, intrigued, posed for five sessions, making thirty 'light drawings' suspended in the air using a small electric light (see Figure 3.8). Multiple cameras and carefully timed stroboscopic flashes in a darkened room allowed the film to capture a still image of the artist and the illuminated lines of the moving light.[19] Picasso appears lithe and energetic; the drawings express his rapid, fluid motions. The motion-captured drawing was invisible to him as he drew, appearing for the first time once the film was developed. Emphasizing the speed of his movements, these photographs reinforced the idea of spontaneous, sudden, creative action, like a flash of light.

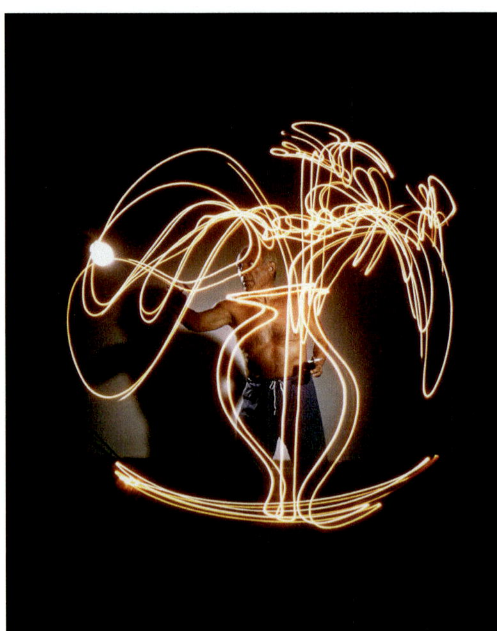

Figure 3.8 Pablo Picasso making light drawing, 1949, Gjon Mili. Shutterstock.com.

But the gestural, bodily line was not exclusively associated with this image of creativity or the optimal body. Frederick Kiesler's (1890–1965) *Endless House* project evolved slowly from the early 1920s until 1960, when a large model was included in the Museum of Modern Art (MoMA) show *Visionary Architecture*. Kiesler imagined a building that would be 'endless like the human body', formed by a new body-centred drawing practice.[20] Rather than being projected up from a floor plan, he wanted to 'project on paper the nucleus of the "dwelling-universe": *the nervous system of the house*'.[21] It was not a matter of composing a house but of feeling out its living form. Endless, like the lives of its occupants, this form was also cyclical. The biomorphic shapes of the *Endless House* are not so much formal references to the shapes of organisms but expressions of this cycling. His drawings feature swiftly drawn arcs, traced and redrawn. He does not seem to be searching for one particular curve but allowing each one to redefine the project. His models, too, are haptic rather than formal, made by bending and plastering. In a well-known photograph of Kiesler at work on a half-scale model for the MoMA exhibition, he is enclosed in a bubble of flexible wire mesh, shaping it by hand from the inside out[22] (see Figure 3.9). He imagined a prehistoric creator who did not plan ahead but 'formed space directly into a house', continually inhabiting, testing and experiencing.[23] On the surface, it seems like Kiesler abandons drawing but it would be more accurate to say he reinvents it as something performed directly in space.

Technologies of motion capture suggest a way to make spaces shaped directly around and for the human body. Imre Makovecz's (1935–2011) *Minimal Space* (1972) mapped motion to define building form. Small light bulbs were attached to the limbs and joints of models, and they were then photographed in the manner of the Gilbreths. An interior was formed by the lines revealed: Edwin Heathcote writes, 'The trajectory of the arm movements were seen as defining an inner shell, or containing body; the curvaceous interior forms, similar to the shapes of the ribs, were each perceived as a film-frame of the movements in one position.'[24] By tracing the movements of the body so closely, architecture could become like clothing, skin or a tightly fitted shell.[25] Where the Gilbreths were concerned with minimal effort, Makovecz sought to express an existential minimum. Coop Himmelb(l)au also saw potential for motion capture. Their *Open House* (1984) was designed by gesture. They filmed themselves against a black background, intuitively gesticulating with their hands and arms to describe the house. Then, with their eyes closed, they made a 'frantic sketch . . . a "psychogram"' on paper.[26] As with Makovecz, the gestural drawing was a moment of free expression at the beginning of a project but needed to be progressively settled and translated into a fixed, communicable form that could be projected towards building.

At the cusp of digital integration within architecture, the gestural drawn line was understood as performative, expressing intuition, emotion and the subconscious. And so, the hand-drawn sketch was placed in opposition to the apparently

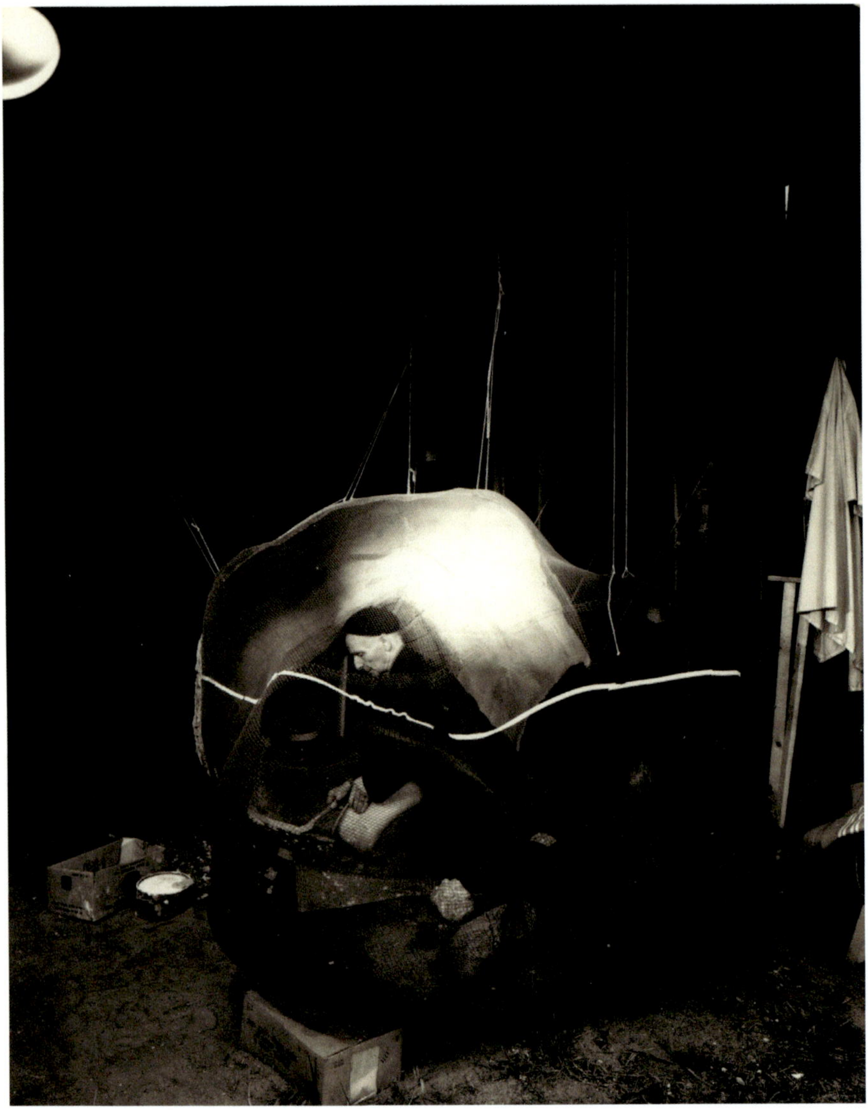

Figure 3.9 Portrait of Frederick Kiesler at work. Black and white photograph. Amagansett *c*. 1965. Photograph by John F. Waggaman. © 2024 Austrian Frederick and Lillian Kiesler Private Foundation, Vienna.

mechanical operation of computational drawing. Frank Gehry's looping, twisting lines, laboriously translated by others into digital models, are signature gestures and assertions of authorial intent. Remaining faithful to the irregularities of his dashed-off drawings required considerable technical expertise. In reasserting a traditional division of architectural labour, and in seeing the freely sketched line as a precursor to a technical realization, Gehry's drawing, although it sought to express movement, largely reinforced traditional ideas of architectural drawing. To draw freehand was haptic – 'It's almost like I'm grinding into the paper, trying to find the building'[27] – a way to know and experience form beyond the visual. The gestural line is a trajectory through time: a journey, not an itinerary. It might express confidence and virtuosity, as in Mili's photographs of Picasso; reveal life and evolution like Kiesler's; or flag hesitancy as it did in the Gilbreths' studies. In all these cases, however, the resultant drawn trace ultimately froze the gestural line. Performative qualities may have been indexed or alluded to but were cast into the past tense.

Architectural drawing similarly may be seen as suspending time or avoiding it altogether. The discipline seems to grapple with time: although entwined, the record of spatio-temporal dialogues demonstrates a provisional relationship. Writer and architect Jeremy Till cites time's uncertainty, its lack of essence, as framing the difficulty of reconciling time within architecture. This, he argues in *Architecture Depends*, is due to architecture's reluctance to admit contingency within its processes.[28] Architectural representation is predominantly concerned with architectural space yet to be materialized. Drawings are produced to picture an imagined building, to assist in 'getting to' the building. Within architectural drawing, intentions are captured – *frozen* it could be said – for a future, as yet unbuilt project. Hence, the dominant practice is in producing drawings whereby the subject matter exists after the drawing, not before it.[29] David Leatherbarrow writes, 'Architects work not in the nominative but in the subjunctive case; each drawing or model is an "as if".'[30] Orthographic projections are not commonly drawn with vigour or spontaneity. Due to their precision, they seem less connected to the physical act of drawing and, hence, the present. Their generative qualities refer strongly to a future tense, proclaiming *this will be*.

The present tense, then, is the time of the drawing's viewer. According to Michael Newman, the drawing's particular mode of being 'lies between the withdrawal of the trace in the mark and the presence of the idea that it prefigures'.[31] As it was refracted through the performative turn of the 1970s, however, drawing was resolutely present-tense. Rather than the production of durable artefacts, drawing was taken to be an event, something that happens in the world. Such works could be documented, or the remains preserved, but these were not themselves the drawing. We look now at architectural drawing in which the present tense is the time of the drawer and so encompasses duration.

Gesturing towards architecture

The idea of choreography provides a line of inquiry towards flux and duration rather than stasis. Instead of expressing movement through unchanging architectural form, the choreographic approach portrayed architecture as a moving arrangement that an architect, like a choreographer, could only partly determine. Allan Wexler's *Crate House* (1990) divided the functional programme of a home into a set of four basic activities (washing, cooking, living, sleeping) (see Figure 3.10). Each of these was served by a collection of associated artefacts in large, wheeled crates; these had no specified relationship to one another, rolling free in a gallery interior. The house would be animated by its occupants: pushed, pulled, opened and repacked. Wexler defined the architecture of *Crate House* as 'choreography without a choreographer, structuring its inhabitant's movements'.[32] If the house is a field of items moving in a kind of performance, as Wexler suggests, the role of the architect is no longer to prefigure its form but to set the stage for its unfolding. The crates were deliberately unsophisticated in their construction and finish and the items they contained were quotidian. Wexler was less interested in *expressing* movement than in fostering actual movement: the house would be performed. Also drawing on scenography and performance, John Hejduk's (1929–2000) *Victims I* (1984) is a proposal for a memorial park in the form of a collection of discrete architectural characters scattered in space (see Figure 3.11). They were more like actors in a play than sets for one: wandering 'vagabonds' rather than residents.[33] The plan provided a provisional arrangement; Hejduk insisted many other configurations would be possible. It was a script or score that could be performed in more than one way.[34] Wexler's

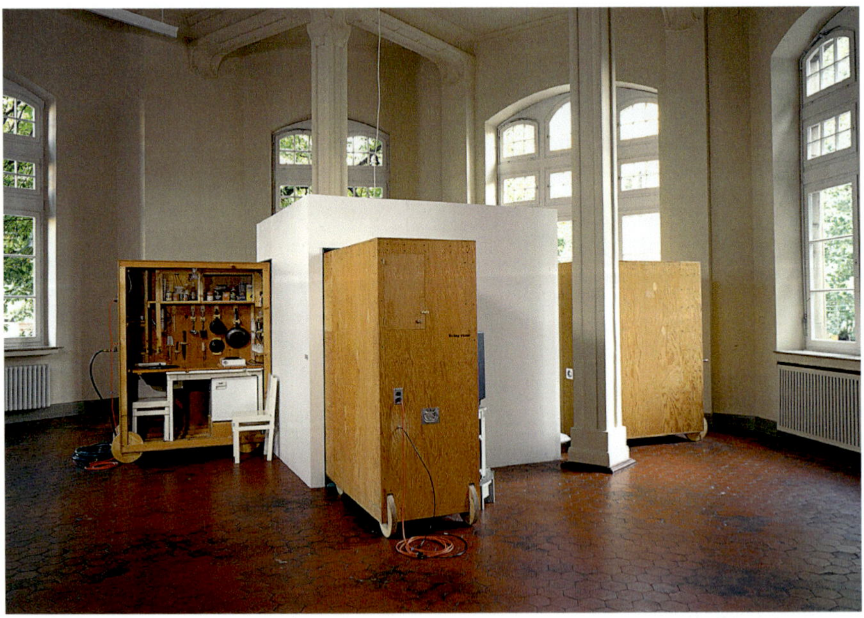

Figure 3.10 *Crate House*, 1990, Allan Wexler. Osthaus Museum, Hagen, Germany. Photograph by Achim Kukulies, Düsseldorf, Germany.

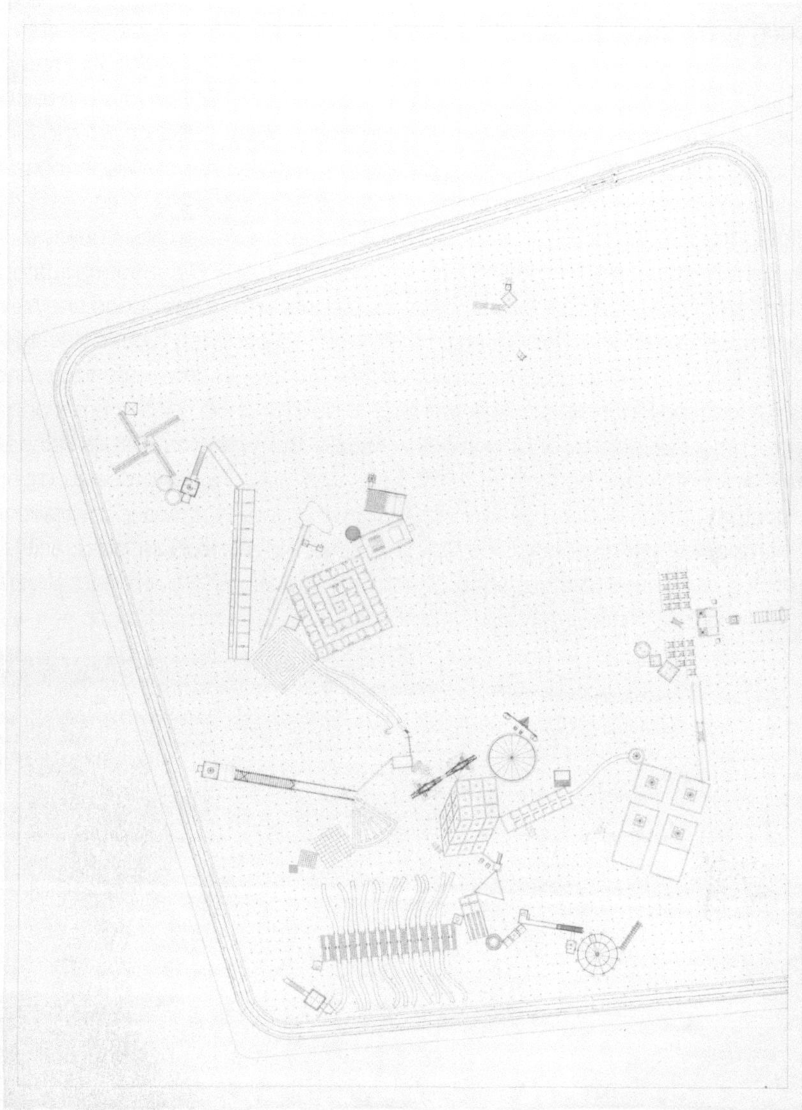

Figure 3.11 Partial site plan for *Victims I*, 1984, John Hejduk. Drawing in ink on paper; 38 25/32 × 53 15/16 in. (98.5 × 137 cm). DR1998:0109:003:017. John Hejduk fonds, Canadian Centre for Architecture © CCA.

and Hejduk's choreography situated architecture in the time of its performance rather than in an originary gesture. Architecture is enacted by movement rather than being prefigured by it. The possibilities of the work are to be discovered through its performance rather than its form.

The performance of labour provides another line of inquiry against stasis. In Chapter 1, we noted the impact of performance and installation on architectural inquiry at Cranbrook Academy of Art. In the 1990s a series of works at Cranbrook refuted the Renaissance distinction between the creative labours of the drawing board in a studio and the menial work of the construction site; they explored the critical and creative possibilities of the physical work of building and demolishing, making and moving and undoing. In 1991, Dan Hoffman, former head of architecture at Cranbrook, cast the labours of a construction site as the re-performance of lines. For *Recording Wall*, he laid concrete blocks 8 feet high and 16 feet long, photographing each of the 150 blocks from both sides as it was placed (see Figure 3.12). The wall was unbuilt, and the photographs printed directly onto the corresponding blocks. Finally, the wall was rebuilt using the image-bearing blocks, becoming, as Hoffman describes, 'a literal recording of . . . the postural effort required to build it'.[35] Construction, Hoffman demonstrated, is an embodied performance. The wall is not just a direct concretion of a line on a drawing but a slow performance of spatial separation. The architect's line on paper, as a notation, is largely agnostic about the specificities of labour, he writes: 'The circumstances of [a built line's] construction are meant to be transparent

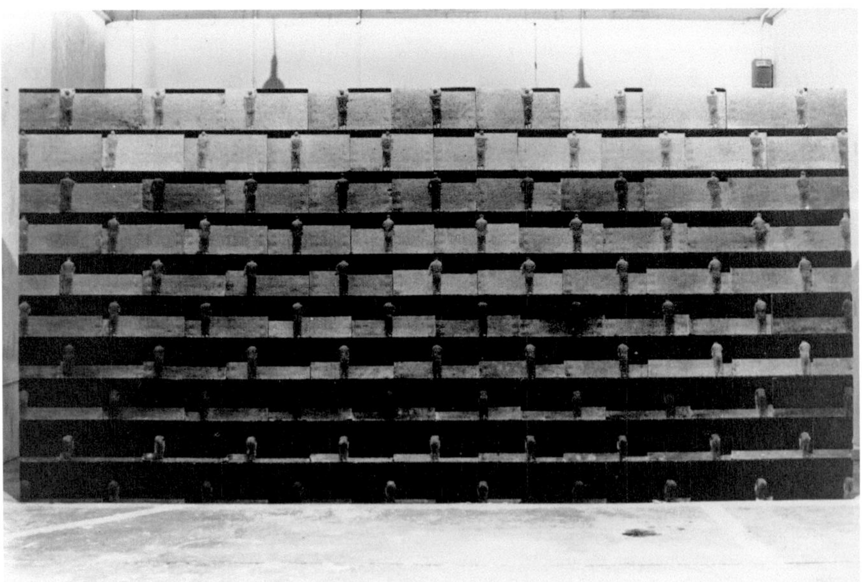

Figure 3.12 *Recording Wall*, 1991, Dan Hoffman.

to the geometric field that is its primary reference.'[36] Such scepticism cannot be maintained in the actual labour of making a real line, however. *Recording Wall* foregrounds tension between the time of the line's construction and the preliminary time of its projection in advance. Laying blockwork is repetitive and involves a disciplined, practised body, diametrically opposed to the free, fluid movements of a sketch, the unrestrained twistings of Allen's painted performers or Schneemann's aerial contortions. A brick wall is a visible record of repeated action, a manifestation of seconds, minutes and hours. Like Brown's *Man Walking Down the Side of a Building*, it bridges the laborious gap between a projected line and an actualized one.

As a kind of critical motion capture of labour, Hoffman's work recalls the efforts of the Gilbreths' subjects. Before chronocyclegraphs, Frank Gilbreth looked for inefficiencies on construction sites, publishing a photographic guide to correct bricklaying.[37] The gap – between the projected line and the performed one – highlighted by these works became more acute with the computerization of architecture and the advent of digital fabrication. Computers encouraged assumptions of extraordinary precision, at odds with the realities of making things. As Francesca Hughes puts it, we plan architecture using 'contemporary CAD software in which we draw brick walls (which we know will be built on muddy sites, by workers wearing thick gloves) to six decimal places'.[38] These pristine, precise drawings then themselves ended up on the building site, handled by many, 'splashed with cement, covered with lime and morning tea, rubbed, doodled on, soaked by the rain and dropped in the mud. By the end of the job the drawings are usually a sad pile of tattered pulp ready for the bin.'[39]

At Cranbrook in the 1990s a series of exercises given in the first months of the programme were set out through problem statements, such as 'Form a Pour' and 'Make a Temporal Section'.[40] These asked students to explore the architectural implications of time for drawing. Another problem statement, 'Body Measure, Measured Body', coupled the fixed and spatial aspects of measurement with the unpredictable and variable characteristics of the body. Jeanine Centuori investigated this through an apparatus in which unconscious bodily movements were recorded through drawing. *Body Measure: Device* was mounted above the head of the wearer with a string connecting hand to pencil; a line is drawn when the hand is moving (see Figure 3.13). The resulting drawings measured movement and, as Hoffman wrote, create a picture of 'a body measure that acts as an index of our own movements in the world'.[41] Conventional architectural representation was inadequate for a 'field of encompassing, environmental concerns', but they could be approached by activating and recording the interior surfaces and spaces of the studio.[42]

Computation also enabled novel ways to bring time into drawing, particularly through a new evolution of the gestural line: flows. Where the sketched gesture or the motion-captured figure draw attention to the movements of an individual

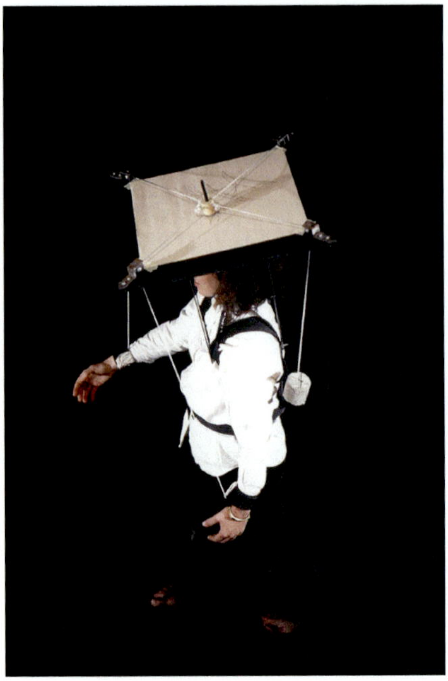

Figure 3.13 *Body Measure: Device*, 1989, Jeanine Centuori. Wood, cables, paper and pencil.

body, flow concerned the movements of populations: meta-gesture. Similar to motion-capture studies, flow diagrams were features of scientific management from the early twentieth century. Human bodies were treated like water or air and represented with directional lines. Arrows and thin lines could be used to express either actual or intended movements in aggregate, and flowcharts became important architectural drawings, potentially usurping the sketch's generative role. By the 1970s, flowcharts were common in computer programming and were on their way to becoming a general-purpose strategy for visualizing all kinds of processes.[43] In architecture, flows came to describe something between form and activity, finding their ultimate expression in the 'parametric' style of the early 2000s and 2010s. Swarms, electrical fields and fluid dynamics were taken to capture mass movement, disclosing a collective unconscious just as the gestural line disclosed the unconscious of the individual.

In the face of industrialized processes, the deliberate expression of time through painstaking or slow methods can be an act of resistance. The Dipdii Textiles project (2018–), initiated by architect Anna Heringer and tailor Veronika Lena Lang in collaboration with Bangladeshi textile workers, 'challenge[s] the linear spatiotemporal character of the usual lifecycle of an object of everyday use'.[44] Each work in the collection is stitched and hand-embroidered (with plans, maps, elevations and words) from worn-out saris, offering 'a possible archaeology of the quotidian, in which traces of previous lives are added to the ones imparted by the process of collective and participatory production'.[45] Rather than registering momentary gestures or confident sweeps of the arm, they index the collective movements of many hands.

These are only some of the ways architects have explored architecture's relationship to the moving body. If gestural lines are understood as preliminary to building, their movement is captured and fixed like a pinned butterfly in a tray. The sketch, whether on paper or described in three dimensions as a motion trace, can be used as a method for form-generation or a means to analyse use. In these, the architectural process is fundamentally about determining the static facts of a building. But if architecture is to engage with change, lifespan, variable conditions and uncertainty, drawings that perpetuate motion instead of freezing it are crucial. Rather than movement that leads towards building, we look for the ways in which movement reveals architecture as an ongoing performance. For this, we turn to architectural drawing practices that *unfix*.

Moving in circles

Drawing the Room | Drawing within the Room (2019) is a full-scale, performative architectural investigation of motion capture's speculative possibilities. Carried out by Sarosh Mulla, Aaron Paterson and Marian Macken, it used captured motion to undertake post-occupancy evaluations of three houses designed by Mulla and Paterson's architectural practice, Pac Studio.[46] The recorded data, as a three-dimensional drawing, was then re-projected into different spaces at full scale, overlaid with the BIM model, creating conjunctions of spaces and times.

Occupants of the three houses – Fyvie House (2015), Space Invader House (2015) and Yolk House (2018) – were recorded as they engaged in everyday household activities such as cooking, eating a meal, running up and down a set of internal stairs and vacuuming. Their movements were captured by a net of six cameras arranged around the spaces being studied (see Figure 3.14). The occupants wore bodysuits with thirty-seven attached tracking markers as consistent reference-points and daylight and reflective surfaces had to be excluded. This was not inconspicuous fly-on-the-wall documentation, nor a spontaneous record. Residences were temporarily converted into motion-capture stages filled with camera tripods, snaking cables and laptops.

The captured videos were processed digitally to triangulate the spatial position of the tracking points. After processing, the raw data yielded animations which were rendered as lines in motion that faded away like afterimages[47] (see Figure 3.15). Occupants' bodies are represented by a cluster of moving lines,

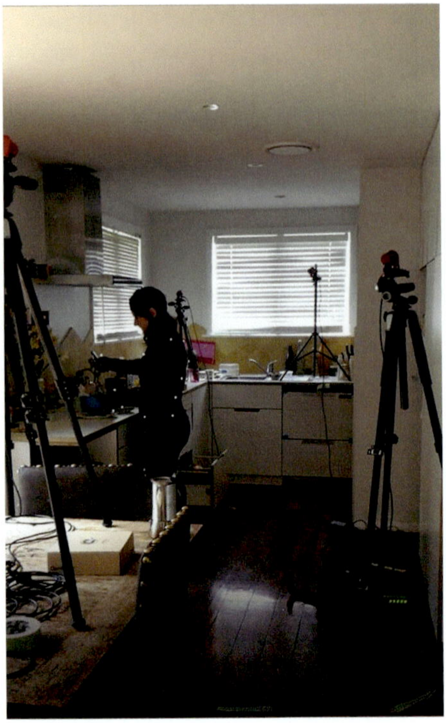

Figure 3.14 *Drawing the Room | Drawing within the Room*: motion capture, recording the act of cooking in Pop Up Gable House, 2019, Sarosh Mulla, Aaron Paterson and Marian Macken. Photograph by Pac Studio.

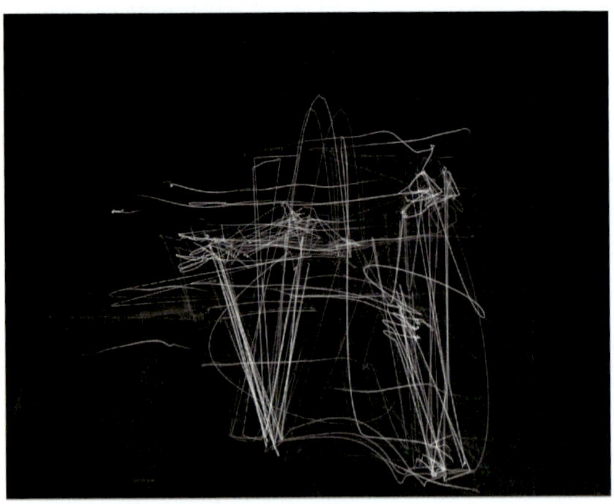

Figure 3.15 *Drawing the Room | Drawing within the Room*: still from motion capture animation, 2019, Sarosh Mulla, Aaron Paterson and Marian Macken.

as spatial constellations rather than solid objects, and the building is rendered immaterial and transparent. The data is visually light: a flock of fine, linear traces expose the essence of occupancy within the ghosted mass of the building: a ghost inhabiting a ghost. As motion-captured traces bound up a staircase, we can see through the walls into a bedroom beyond. The material presence of the building returns through the recorded audio of the scene. Footsteps on a hard floor and clinking crockery in a sink conjure familiar scenes for us, prompting us to interpret the traces and aural atmospheric clues to provide spatial form. The sample rate of the capture smooths out irregularities. If, like the Gilbreths, we see motion-captured lines as revealing psychology and expertise, we might read these animations as disclosing the practiced confidence of familiarity. But where the Gilbreths' recordings freeze motion to study it, the looping format helps us imagine ongoing, habitual moments. These motion traces are not prompts for optimization but a way to confront the unruly animacy of human bodies. The work was exhibited as full-scale projections in Pac Studio's architectural office, with locations chosen to coincide with the activity depicted.[48] The act of cooking documented in the Fyvie House was superimposed with the space of the office kitchen; the recording of eating a meal at the Yolk House was projected over the communal space of the long office meeting table; and a projection of someone running up stairs was situated to create the illusion of viewing through a thin curtain (see Figures 3.16 and 3.17). Movement tethers two spaces. The

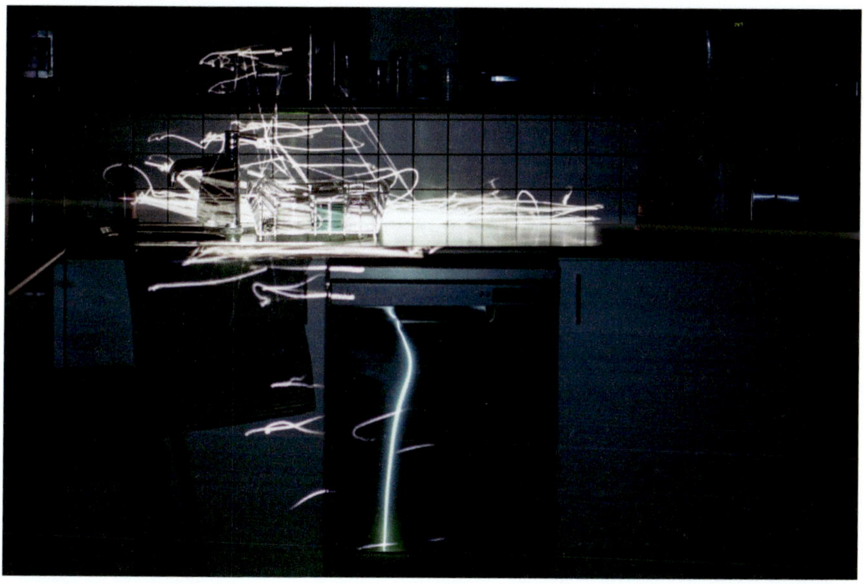

Figure 3.16 *Drawing the Room | Drawing within the Room*, 2019, Sarosh Mulla, Aaron Paterson and Marian Macken. Photograph by David St George.

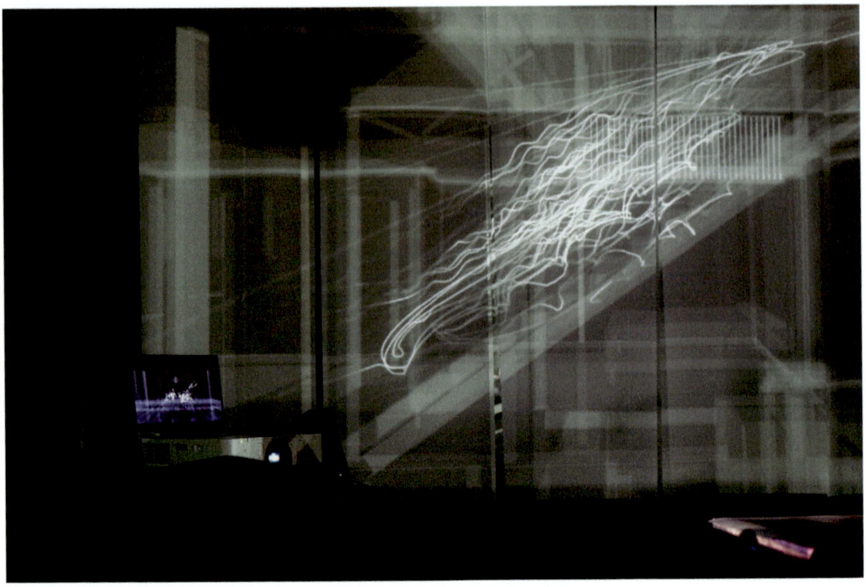

Figure 3.17 *Drawing the Room | Drawing within the Room*, 2019, Sarosh Mulla, Aaron Paterson and Marian Macken. Photograph by David St George.

spatialized drawing allows the viewer to be positioned at a place between multiple worlds. Viewers watching the recorded activity playing out in the dimmed studio could not help casting their own shadows into the image, temporarily becoming the most solid thing in the projected image.

Architectural drawings used projectively to anticipate future construction tend to abstract and generalize use and occupancy, fixating on the permanent and omitting the temporary. Post-occupancy studies reveal a different view, highlighting the ways that people appropriate and make use of spaces in their own temporary ways. Pacific academic Karamia Müller, reflecting on her own 'relational' post-occupancy study drawings, suggests inherited drawing conventions and tools '[do] not address the Pacific view of connections in space that are located in the relational rather than distinct enclosed spaces'.[49] New drawing strategies are necessary. In her drawings of how Samoans in Aotearoa New Zealand make use of 'ancillary diasporic structures' like garages and porches, laundry baskets, shoes, cushions and mats are drawn as specific and significant architectural elements rather than generic entourage.[50] Like Mulla, Paterson and Macken, Müller looks for modes of architectural drawing that respond to actual bodily activity.

Bodily activity becomes a way for separate projects to interrelate. Mulla, Paterson and Macken made other drawings, overlaying plan views of the motion capture within the houses with the office space (see Figure 3.18). The drawings

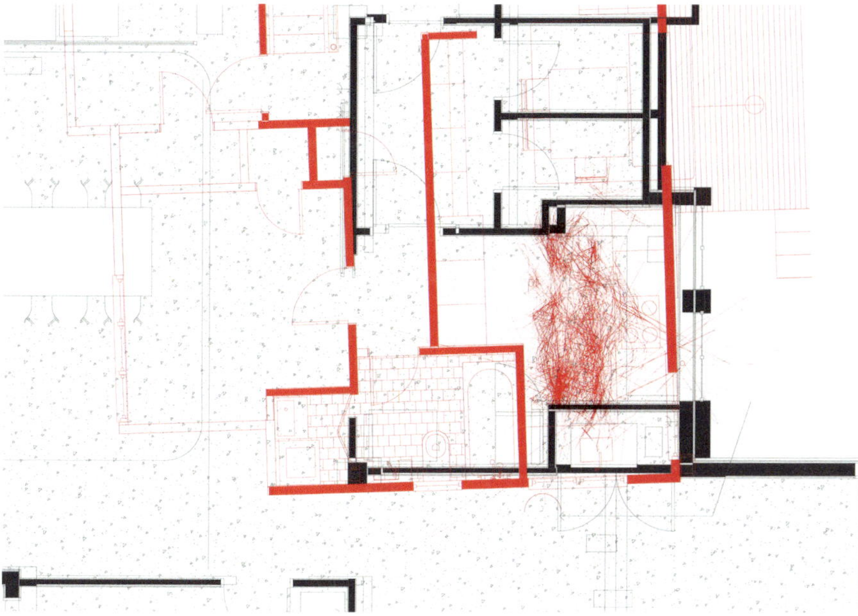

Figure 3.18 *Drawing the Room | Drawing within the Room*: superimposed plans of Pac Studio office and Pop Up Gable House with the trace overlay of the inhabitant's movement while making an omelette, as line work animation from motion capture, 2019, Sarosh Mulla, Aaron Paterson and Marian Macken.

and projections provide a means to see between and across projects. Although the occupants of the houses experience them singly and separately, the architects look across them to see the repetitions and differences of an ongoing practice. The linear movement between site and studio is muddled as the lives of occupants return to inhabit the office. Working at full scale, they write, 'allows for a merging of the architects' and owners' experiences in their parallel inhabitations and for the designers to examine built works simultaneously'.[51]

The studio is a place of abstraction and ongoing reflection in which even completed buildings remain open to inquiry and investigation. The underlying logic is that completed buildings speak into an ongoing conversation that informs future projects and possibilities. Reflection on a body of practice is enabled by simultaneity; comparisons come into view through difference and parallels: 'The legacy of an office's prior built work invisibly inhabits the space of the office. Architects are always drawing on and referencing prior works: the memory of having been within them informs decision-making toward unfurling designs.'[52] The architectural office becomes at once a site of knowledge construction, an exhibition site and an in situ archive. In their experimental post-occupancy studies, motion capture is used to animate and keep projects alive in the context of a practice, rather than suspending them in service of a project.

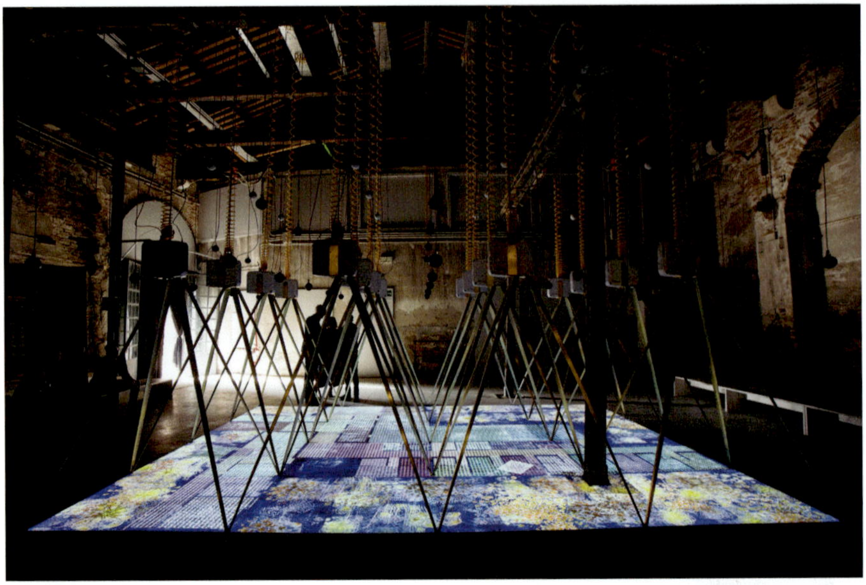

Figure 3.19 *Losing Myself*, 15th International Architecture Exhibition of the Venice Biennale 2016, Níall McLaughlin Architects and Yeoryia Manolopolou. Photograph by Riccardo Tosetto.

Themes of fluidity, suspension and indistinct forms of subjectivity are also key to *Losing Myself* (2016), an animated drawing at the 15th International Architecture Exhibition of the Venice Biennale[53] (see Figure 3.19). This drawing-installation, by Níall McLaughlin Architects and Yeoryia Manolopoulou, explores the spatial experience of people living with Alzheimer's disease and dementia and is based on the previously completed Orchard Respite Centre in Dublin. Like *Drawing the Room | Drawing within the Room*, it is a post-occupancy study revealing the complexities of inhabitation, rather than evaluating from the perspective of optimization. In particular, McLaughlin and Manolopoulou use it to open up questions about the spatial experience of dementia, memory and orientation more generally and the nature of the architectural plan.

Losing Myself consisted of an array of sixteen suspended, digital projectors pointing downwards onto a horizontal screen (see Figure 3.20). Together, they produce a composite animated plan of the Orchard Respite Centre. Wavering freehand lines, elaborating on the plan, are drawn by a host of ghostly hands. Colours and patterns shift as the drawing cycles between night and day, overlaid by a slower seasonal change. These various temporal rhythms are layered and scaled into a sixteen-minute loop. The visitor might enter or leave at any point, so the work cannot be said to have a beginning or end.

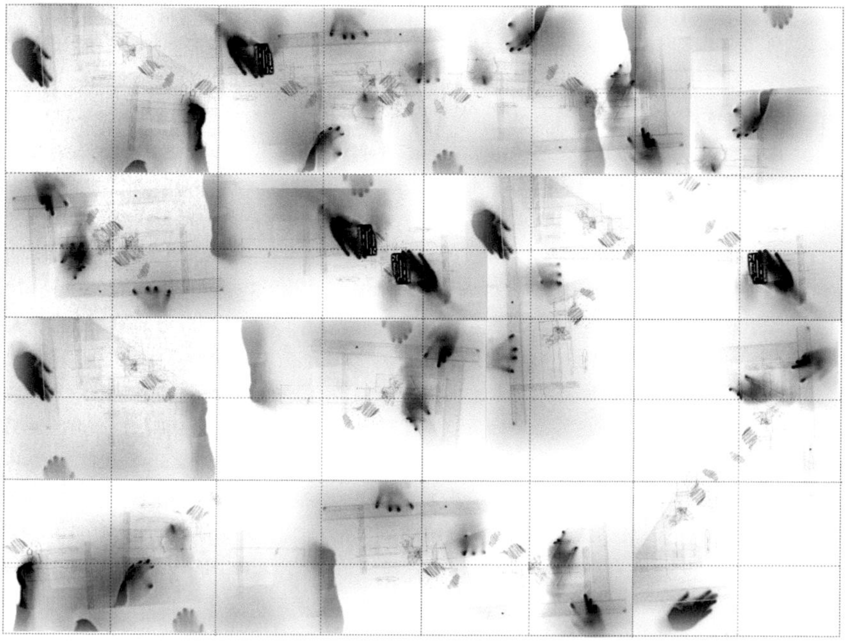

Figure 3.20 *Losing Myself: Polyphony* (video still), 2016, Níall McLaughlin Architects and Yeoryia Manolopolou.

The drawings were made by a group of architects working on drafting film on glass-topped drawing tables, filmed from below. After a series of conversations about the experiences of dementia and Alzheimer's, they were instructed to draw, using a single line, how a person with one of these conditions might experience space. The single-line rule means that various types of drawing merge: a line might track a path through space, then form a pictorial depiction of a potted plant, then indicate the arc of a door swing, plot the thickness of a wall or record the pattern of a surface. The continuity of the line implied the coexistence of these various architectural readings as they were revealed through time. This plasticity was enforced by insisting lines be continuous performances rather than projections leaping across the surface. Sometimes the drawers worked together, 'performing a social situation'.[54] The final animation consisted of multiple, partial views composited together; many hands draw at once. Sometimes the drawing swells with collective activity and at others, settles into stillness.

Manolopoulou adopts a neuro-biological distinction between 'egocentric' and 'allocentric' representation.[55] In the former, the body is a consistent reference point, an 'exact position . . . from which I construct a spatial understanding of the world in front of me'.[56] This mental representation of space in relation to a fixed position corresponding to the self develops early in childhood. In allocentric representation, 'the viewer comprehends the world synthetically by assessing object-to-object relationships without having to physically shift their body to different locations'.[57] Architectural drawing and the imaginative inhabitation of architects requires entering the drawing egocentrically, envisioning it from a particular location and moving around; this also requires slipping into an allocentric mode, grasping unseen relationships. For those living with dementia and Alzheimer's, allocentric understanding dissipates, leaving the individual in an egocentric world.[58]

The many hands of the drawing, the deliberate disruption of architectural drawing's allocentric norms by the insistence on drawing with a single line and the looped presentation, mean that the drawing remains in the present tense. We are not invited to see through the drawing to another building but to watch the drawing being performed. The installation requires the viewer to trace their own lines of movement. The horizontal screen was suspended from brass 'quadpods' attached to digital projectors and hung from the exhibition hall's trusses. The long-tapered legs of the quadpods recall survey tripods or dividers. They stand astride the drawing in perfect array, each with an optical eye staring down between their own legs. Rather than a neat, perspectival window or an integrated orthogonal space, the plane of the drawing is a composite of many shifting views.

The drawing room in which *Losing Myself* was delineated and recorded is a full-scale meta-site, like Pac Studio's office. Strategies of superimposition and simultaneity connect projects to elaborate on a multiplicity of experiences. In both cases, the linearity of the architectural drawing is supplanted by circularity. Notionally complete buildings return to the studio as objects of inquiry. The act of drawing then, as Carole Lévesque and Thomas-Bernard Kennif suggest, 'cycles back on itself in ways that challenge initial intentions and premises, that destabilize and invite uncertainty and skepticism as productive positions'.[59] As Manolopoulou and McLaughlin use it, the performed drawing provides a way to engage with the coincidental and contingent aspects of inhabitation.

If *Losing Myself* leaned into the variability of egocentric human experience, motion capture can also access non-human perception. As we come to rely on computer-mediated vision and computer-controlled action, we increasingly operate through interfaces that capture and convert motion. Vision is no longer a human privilege, but an ambient, technical condition. In *Dream Life of Driverless Cars* (2015), experimental architectural imagers Matthew Shaw and William Trossell (ScanLAB) reveal an uncanny version of the city of London[60] (see Figure 3.21). Marrying laser-scanning with motion-tracking, they mounted a scanner to the roof of a car driven through London. The result renders solid surfaces of the city to appear as superficial veils, full of gaps. Sometimes, there are enough visual clues to make the view familiar, but at other moments, the virtual camera moves impossibly, or objects remain unrecognizable: Big Ben appears in two places at once and a strangely attenuated object is revealed

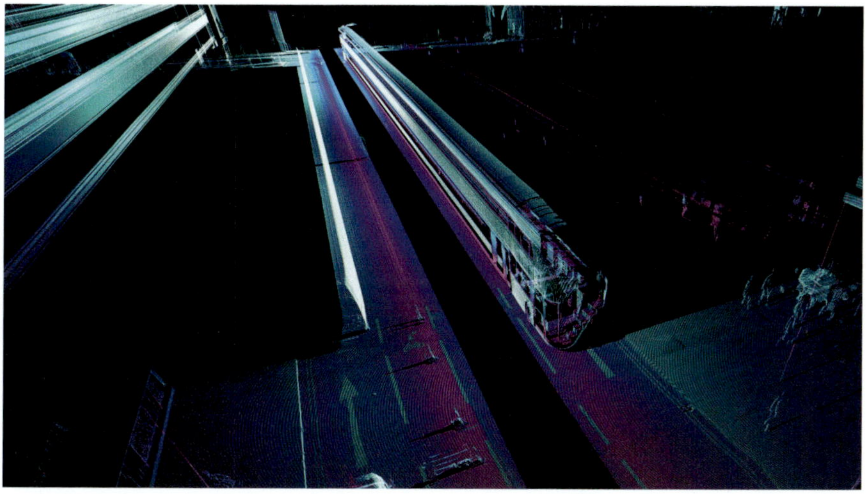

Figure 3.21 *Dream Life of Driverless Cars – The Bus*, 2015, ScanLAB Projects.

to be a bus, distorted as it followed beside the scanning moving car for a distance. Scanners can image motion with extraordinary precision, but motion also prompts hallucinatory effects, reminiscent of the perceptual experience of dementia. Time warps space. In Chapter 2, we looked at the logic of three-dimensional scanning, which depends on projecting lines from a fixed point. The ability to reconstruct a three-dimensional space from a radial scan depends on a fixed point from which the scan is made; the random saccades of the eye are replaced with systematic sampling. Once this point is set in motion to become a trajectory through space, and the objects being scanned are also mobile, the scan produces something aleatoric and dream-like. Shaw and Trossel liken this to peripheral vision, or 'mad machine hallucinations';[61] media theorist Jussi Parikka calls it 'over-seeing'.[62] As the conditions of dementia and Alzheimer's serve to disrupt the allocentric view of the plan in *Losing Myself*, ScanLAB's digital hallucinations disrupt assumptions about the certainty of computer vision. Whether or not the animation itself qualifies as a drawing, it is produced through a performative act of drawing: a line traced through the city with a spinning, linear, rangefinder. The city, like Pac Studio's office and *Losing Myself*'s drawing room, is a meta-site: an open-ended place to which drawings return.

Circular timelines

It should be clear by now that our expanded use of 'drawing' as a verb remains diffuse as to what constitutes 'a drawing' (as a noun). The legacy of the 1970s for drawing is different spatio-temporal understandings, refracted through the new body-centred genres of installation and performance art. Bodies shifted from being thought of as abstract to becoming inescapably concrete: gendered, varied, cultured, socialized and organic. As Schneemann hangs naked in her harness, scratching at the walls and floor, her *inability* to draw is as important as the marks she does manage to make. In this performative turn, the *act* of drawing presides over the drawn artefact, the result being that a drawing process may not necessarily result in something easily recognizable as a drawing. There are obvious challenges in applying these ideas of drawing to architecture. Academic and architect Mark Rakatansky points out that architecture tends to reinforce assumptions about the body:

> What gesture *does*, what gesture *can* architecture make in response to the gestures of bodies? The gestures that conventional architecture tends to make towards bodies is to reduce the body to a series of standardised measures, proper behaviours, ergonomic formulations. In other words, architecture's gesture is to match, and thus spatially constitute, these already determined socially determined reductions of the body.[63]

Architectural drawings – in their traditional, professional and prevailing uses – are disconnected from both the space and time of their production, the site of their enactment and the bodies of those who make them. The drawing passes between studio and site but belongs to neither. Representations – as premonitions of buildings yet to come – are interstitial, enabling mobility, translation, repetition and verification to the extent that they can transcend their enactment in the present.[64] The devices of scale, notation and geometric abstraction enable drawings to navigate and mediate across this separation. Such drawings are suspended in time; or perhaps, as John Hejduk writes in reference to plans, they show 'architecture in the state of sleep. Plans are *sleeping architecture* that, in extreme, are *architecture in death*. We tend not to want to disturb architectural plans, for they are so *still* and so *quiet*. The plan returns architecture to a state of timelessness.'[65] The four acts of drawing with which we opened this chapter are, for us, provocations. What happens if we keep drawing awake? If we refuse to let it slip off to other places and times but instead keep it full scale and in the present? What architectural meaning, use or value ensues? From these questions, we consider performative drawing in the gestural sketch and motion capture. These reveal qualities of lines that are temporal, indexing unconscious, expressive, emotive, intuitive or collective bodies. Whether taken to express a drawer's spontaneous virtuosity (Mili's images of Picasso) or a labourer's hesitancy (the Gilbreths' chronocyclegraphs), gesture is conceptualized as a moving spatial line. These act as *traces* of an *act of drawing* rather than precursors to a drawn artefact.

The gestural line remained a fascination for architects looking to reinforce the expressive, temporal, inhabited qualities of architecture over its abstract, rational ones. In doing so, however, the assumption that the gestural line eventually crystallizes into fixed forms of building largely went unquestioned. The notion of choreography began to suggest an alternative route, by conceiving of an architecture that remained in motion. Hoffman's *Recording Wall* (like *9119 St. Cyril* in Chapter 1) folded the linear relationship, between image and building, back onto itself. Our contemporary examples – *Drawing the Room*, *Losing Myself* and *Dream Life of Driverless Cars* – draw *from* architecture rather than *towards* it. In doing so, they offer at least two significant opportunities for architectural inquiry: they invite experimentation with time as corporeal and plastic, and they provide resources for understanding questions of site-specificity.

Marco Frascari (1945–2013) writes of architectural time, not as a series of present moments, but as 'temporal "extantness" that make past, present, and future coexist'.[66] Time for architecture is multi-layered but the drawing/building split enforces a more limited temporality: projection towards the future. Performative drawings, by addressing the body, engage a more complex sense of corporeal time. As interdisciplinary scholar Maxine Sheets-Johnstone writes, 'We do not . . . experience ourselves and one another as "packaged" but as moving and moved, in ongoing response – that is in correspondence – with the things around us.'[67] Time, in this sense, is not a fourth dimension to be added to the three spatial dimensions of Cartesian–Newtonian space. Rather, it is oriented by the possibilities and relationships offered to us by our environments, conditioned by the potentials and limits of our bodies and entangled with our imaginative investments in the world. 'In a kind of double moment', architectural researcher David Turnbull writes, 'people perform objects of all kinds, but especially buildings, by moving through and around them but buildings also perform people by constraining their movements and by making likely certain kinds of encounters between them and others'.[68] Drawing from architecture – tracing its occupation, retrospectively inhabiting it through drawing or re-imaging it – gives us the opportunity to investigate this mutual performance. The result is that we start to understand architecture similar to how theoretical physicist Carlo Rovelli writes about our existence more broadly: 'We can think of the world as made up of *things*. Of *substances*. Of *entities*. Of something that *is*. Or we can think of it as made up of *events*. Of *happenings*. Of *processes*. Of something that *occurs*. Something that does not last, and that undergoes continual transformation, that is not permanent in time.'[69] Just as the drawn line is in constant motion, so too are buildings continually under construction, being made; scholar of sacred architecture Lindsay Jones writes that buildings 'are never "done" in the sense of fixing and fully stabilizing their meanings. . . . The idea of the fully finished building is a fiction that cannot be sustained.'[70]

Ideas of site and site-specificity, argues art historian Miwon Kwon, are affected by contemporary art's shift of focus from nouns (object and artefacts) to verbs (processes).[71] Hence, the relationship between a work and its 'site' is no longer based on physical permanence but rather on the 'recognition of its unfixed *impermanence*, to be experienced as an unrepeatable and fleeting situation'.[72] Practices of performative drawing shift the location of drawing; for artists, places of creation move from studios and galleries and into public spaces of landscapes and cities. For performative architectural drawing, the notion of site changes. An architectural site usually refers to the location for the *content* of a drawing, not the location of the *act* of drawing. While an office's location is recorded as a component of the documentation set, this does not reflect the importance of the place of drawings' production. Performative architectural drawing explicitly activates the place of drawing. Practices that merge the sites of architectural production and drawing with inhabitation and built outcomes address the specifics of relations; that is, as Kwon writes, 'the uneven conditions of adjacencies and distances *between* one thing, one person, one place, one thought, one fragment *next* to another, rather than invoking equivalencies via one thing *after* another'.[73] In the following chapter we will consider how this shift from sequence to simultaneity changes our sense of what it might mean to *inhabit* drawing spaces.

4
INHABITING DRAWING
FIELDS OF ENCOUNTER

Exposing drawing

Exhibitions have become essential sites for contemporary architectural practice. Prestigious international events like the Venice Biennale of Architecture (established 1980), galleries like Storefront for Art and Architecture (established 1982) and collections focused on architecture like the Canadian Centre for Architecture (established 1979) built on new possibilities identified in the late twentieth century (see Chapter 1). Since then, these sites have multiplied. As well as venues for (self-)promotion, they are a means to publicize disciplinary concerns and situate practices within wider socio-political and cultural contexts. Crucially, exhibitions offer a certain detachment from professional imperatives: architectural researcher Thordis Arrhenius writes, 'Unhinged from the political, ideological, and economic forces conditioning architectural production, the exhibition provided a space in which these forces could be displayed, their implications revealed, and their trajectories acted out.'[1] Exhibitions coordinate professional attention and engage publics, producing narratives and meanings across or between projects. As well as representing or translating architecture carried out elsewhere, they have become sites to test, prototype, model, speculate or criticize. But exhibitions pose particular challenges for architecture, some of which we note earlier: there is a tension between showing architecture and the architecture of showing; and the exhibition of architecture always involves some degree of displacement or indirectness. A newly spatial and performative approach to drawing, however, activated novel architectural possibilities for exposition. As Rolf Hughes puts it:

> Exposition has become a key term for distinguishing artistic research from artistic practice. Put simply, it is the strategy or set of strategies by which the researcher stages an encounter with artistic research such that the contribution becomes implicit or explicit. But an exposition is not an illustration of an idea or theory. It is rather the creation of fields of encounter.[2]

In this chapter, we look at a series of recent full-scale drawings carried out as creative practice research inquiries in academic contexts. In the rising academic paradigm of creative-practice research, exhibition is an expository site of dissemination, appraisal and critique. Exhibitions of architecture – through historical surveys, the making and display of proto-architectural objects and speculative design schemes and digitally mediated installations and environments – critically investigate space and the politics of place and so are key forms of research inquiry and output in this field. The works in this chapter do not describe architecture situated elsewhere but enact it by creating *fields of encounter*. We take this term to mean more than didactically showing or descriptively revealing to encompass relational fields activated by the physical presence of the viewer. Earlier chapters have mainly considered drawing from the perspective of the drawer, emphasizing the performative aspects of *making* drawings. In this chapter, we bring viewers' bodies to the discussion to focus on *encountering* drawing. To speak of a *viewer* is of course to privilege the visual. Here, however, we intend that it be understood as a placeholder term for the various ways that bodies encounter drawing: as perceivers, interpreters, readers, participants, audience or otherwise affected parties. The drawers themselves may also be viewers to the extent that they too encounter their work in making it.

Our intention is to demonstrate the contemporary vibrancy and critical interest of full-scale architectural drawing; to highlight themes and questions that animate and connect these works, particularly the idea of inhabitation; and to explore the ways they activate the bodily presence of the viewer. The works reviewed here manifest as situated, temporally present, immediately apprehendable constructions in space. They decouple a fixed scaled reading of drawings: rather than bringing together proposed space with existing, they tether and hinge operations between scales, virtual and physical spaces and times. We examine how these drawing-installations approach immersion, articulate relationships with viewers and engage bodies; and how they simultaneously allude to, and layer, other places and times by means of referential conventions and expectations of drawings. In doing so, they join other expanded architectural drawing practices in reevaluating drawing's relationship with real space contributing to occupiable demonstrations of thinking through drawing.

Fields of encounter

Exhibition and installation are not the same thing, yet they have in common impermanence. While the distinctions between the two are not always sharply marked, broadly it may be seen that an exhibition, as a frame around a work or works, is designed in a way that is sympathetic (or antagonistic) to the works it contains; it is conventionally understood to have curatorial intention layered onto or wrapped around it.[3] An installation, in comparison, is more singular even if it consists of many elements and may be described as an environment, scene, relational field or experience. To the extent that an installation concerns time, as well as space, it may be understood as an event or happening. From our perspective, the most interesting aspects of the exhibitions we attend to are that, by virtue of their architectural concerns, they begin to slide towards installation. Once architectural drawings and models are seen to *be* architecture rather than just *refer* to it, they can no longer be simply *displayed*.[4]

Art historian and critic Claire Bishop proposes that installation art is defined by the physical presence of the viewer who is 'integral to the completion of the work'.[5] Installation is immersive; the viewer in some way enters into the work, activates it or is activated by it in 'bodily appreciable' ways.[6] Understandably, phenomenology has often been a resource for comprehending installation. Maurice Merleau-Ponty (1908–61) argues that our physical, living bodies are the ground of experience and thought, and Bishop points to 1960s English translations of his writing as crucial for the development of installation art.[7] These texts argued that experience was not the meeting of a subject with an object, but a relational field within which subject and object were only provisional effects: 'I do not see [space] according to its exterior envelope; I live it from the inside; I am immersed in it. After all, the world is all around me, not in front of me.'[8] This is necessarily an open-ended experience; an installation does not define a singular experience but a field of possible experiences.[9]

Installation art has often claimed site-specificity. The installation is taken not only to relate inextricably to the viewer but also to its context. Miwon Kwon, in her genealogy of site-specificity in art, shows how the concept expands. At first, site is taken to mean 'an agglomeration of the actual physical attributes of a particular location', including the various properties of the architectural setting.[10] It then takes on wider contexts as 'a relay or network of interrelated spaces and economies' before further dilating to encompass 'cultural, social, and discursive fields'.[11] In the first sense, a site might be a room or landscape; in the second, a museum or art market; and in the third, a discourse, cause or debate. Site-specific works, it is increasingly understood, not only respond to or reveal sites but also construct and deconstruct them. As Mark Wigley has written, the architectural site is an effect of architecture, not simply a place where it is located:

> The project is never simply a view of disruption or reinforcement of a site. It projects a site rather than projects onto or from it. Architectural projects are projectors, mechanisms for constructing the effect of site, the effect of that which supposedly precedes them, an effect that is never more than an image, an image that, like all images, can be occupied.[12]

Like experience, site is not one thing but a constantly recirculating 'problem-idea'.[13] Experience and site, as aspects of installation, define fields of encounter that are open-ended and organized relationally. These fields are further complicated when they specifically engage with architectural drawing. In architecture, the surface of a drawing is an interface between two sites: the studio, or wherever drawing is physically enacted, and the construction site, where building is enacted. Drawings mediate between these two sites and situated actions. They are transitive; as Sophia Banou writes, 'meaning is not simply produced but excavated out of the real . . . through a recursive process of inscription and reiteration'.[14] They are both here and already elsewhere; we noticed this difference in Chapter 1. Architectural drawings, even when considered as artefacts independent of the process of building production, remain ambiguously sited. The works we now address exploit this ambiguity and articulate Banou's description of 'inhabitable places that invite and evoke the engagement of the viewer . . . by defining physical environments but also by opening up these environments to wider and larger situations and spaces of representation'.[15] Full-scale architectural drawing-installations are both *here*, inhabitable with our whole bodies, and *elsewhere*, occupied through a projective imagination.

Ambiguous inhabitation

A poetic story of an immersive, full-sized drawing of Baghdad dramatizes the idea of inhabiting drawings. In 754, after his brother's death, Abu Ja'afar was acclaimed as al-Mansur, the victorious Abbasid caliph. He set out to establish a new capital and the site of the village of Baghdad was chosen, an existing meeting place of caravan routes in a fertile plain at the joining of the Sarat Canal with the Tigris River. He envisioned a circular city with his palace and mosque at its centre, and brought together engineers, builders and surveyors from the empire for the purpose.[16] The medieval Muslim scholar al-Tabarī (839–923 CE) recounts how al-Mansur visualized this proposed city:

> He then proceeded to enter through each gate and walk among its outside walls, its arched areas, and its courtyards, all of which were outlined in ashes. . . . Having done that, he ordered cotton seeds placed on this outline and oil poured on it. Then he watched as the fire flared up, seeing the city as a whole and recognizing its full plan. Subsequently, he ordered the foundations to be excavated along those lines and commenced its construction.[17]

Al-Mansur experienced this gloriously ephemeral, burning plan of proposed spaces at the site of intended construction. In this scenario, the viewer is housed within a full-scale drawing on site, similar to being within survey set-out lines on a newly begun construction site. The flaming full-scale drawing exposes the order of the city that will be concealed once it is built. The city is burned before it is built, perhaps suggesting purification. The plan is animated twice: al-Mansur walks the plan on the ground and then stands back and watches it flare to fleeting life.

Contemporary virtual reality (VR) software is used by many architects in a similar manner to communicate proposed designs to clients who may be more accustomed to filmic special effects than reading architectural drawings. Given orthographic representation's perceived inaccessibility to untrained eyes, VR walk-throughs of proposed designs enable comprehension of spatial intentions. In these scenarios, the client experiences the proposed design, moving between rooms and seeing options for interiors. The assumption is that the VR model provides a direct, transparent view into the future: a simulation. David Ross Scheer claims that the shift towards computational design has resulted in the assumption that reality will conform to the simulation, and that the aspects simulated – mainly visual ones – are the ones that matter most.[18] In the process, the body is set aside. It is not true sunlight we sense on our skin as it streams into these living spaces but rather simulations of light patterns.

A similar client-oriented technology is provided by the company Lifesize Plans. Using integrated projection technology in warehouse-size facilities, clients are invited to wander through plans of their proposed house, augmented by moveable, real furniture[19] (see Figure 4.1). Unlike VR, the clients are invited to use their physical bodies to move and mime their future inhabitation. The experience is shared: a group can move through together, gesturing and seeing one another in space. The projection system facilitates, for the untrained, the kind of imaginative inhabitation that architects habitually employ as they draw. As with the prevailing uses of VR in architecture, the emphasis is on closing the distance between representation and projected reality, to remove ambiguity and uncertainty so the client knows what they are getting.

A contrast to this gap-closing is provided by *On the Surface* (2014), an architectural drawing-installation by architecture and urbanism studio Metis (Mark Dorrian and Adrian Hawker). A large carpet-drawing (16 × 10.5 m) provides a

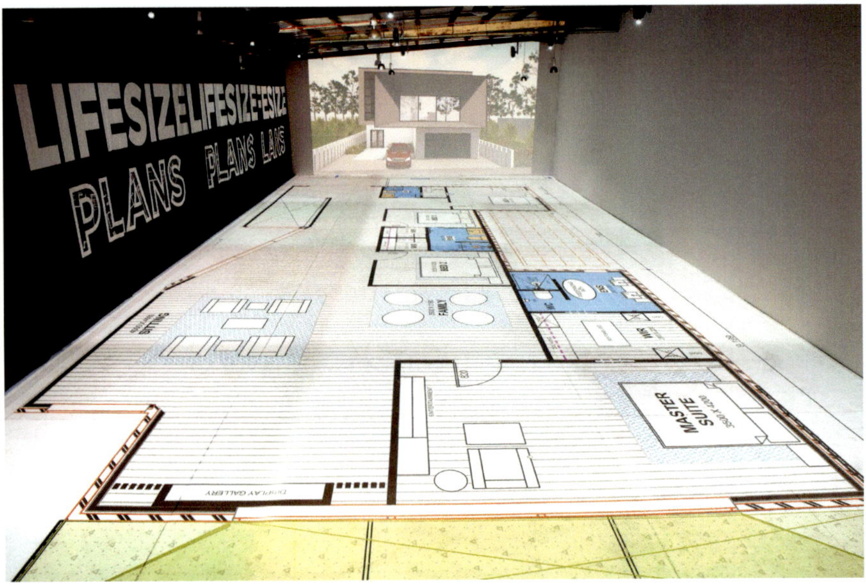

Figure 4.1 LifeSize Plans Sydney – Showroom. https://www.lifesizeplans.com/blog/lifesize-plans-eyeing-out-us-expansion.

conceptual and propositional terrain, a 'composite cartography of projects and places'[20] (see Figure 4.2). Seven glass-topped trestles carry models; further drawings relating to speculative projects by Metis are sandwiched in the layers of glass. The glass surfaces create a transparent horizon, suggesting screens or vitrines. Each trestle table positions the viewer in a particular location and orientation which create spatial relationships between the table's material and the floor drawing. Installed in a gallery with an upper balcony, viewers from above can observe others' interactions with the exhibit (see Figure 4.3). Viewers walk across a drawn landscape and, in doing so, reference the notion of our miniature selves travelling across scaled plans. Here, though, it is as if the body has become smaller, walking on a table at Metis's studio with their drawings laid out upon it, and the drawings vary in scale and format. The goal is not to provide a sense of immediacy, as it is for Lifesize Plans, but for the viewer's body to be out of scale: in several places and at several sizes all at once. As an installation, *On the Surface* displaces as much as it places. A key reference for the work is Jonathan Swift's *Gulliver's Travels* (1726), that narrative of displacement and variable scale. Like Gulliver looking over Lilliput, or al-Mansur over his burning city, scaled plans and models offer us 'fantasies of pre-eminence'.[21] Rather than treating the viewer's body as a fixed reference, Dorrian and Hawker exploit the medium of installation to decentre it. *On the Surface* is an exhibition displaying a collection of discrete architectural works, but the installation integrates these in a similar manner to

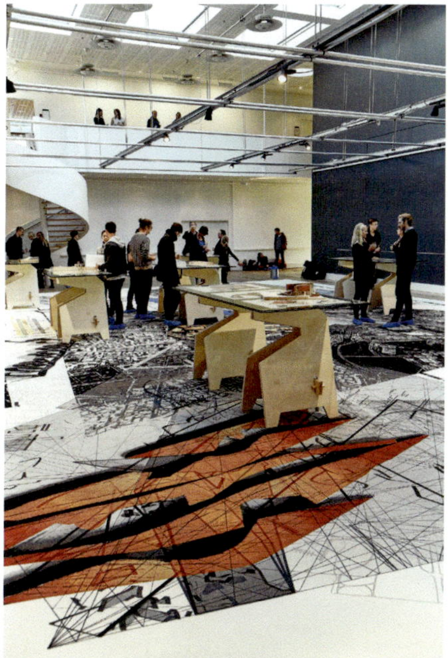

Figure 4.2 *On the Surface*: view with Egyptian Museum project in foreground, 2014, Metis: Mark Dorrian + Adrian Hawker. Photograph by Gert Skærlund Andersen.

collage, treating them as a field of fragments to be assembled by the viewer. Each fragment has its own scale but its differently scaled neighbours suggest that it might be read otherwise.

Installation may be thought to dispense with scale altogether, but it also offers opportunities to think of scale in new ways. A series of installation-drawings by artist and academic Sue Gallagher explores scale as a kind of orientation related to repetition and materiality. The series was initiated with *1 line-dwelling* (2016), a single forged silver bracelet; progressed to *10 line-dwelling* (2017), in which the closed loop of the bracelet was opened and stretched into a dotted line of ten forged hairpins installed on a gallery wall; and then to *100 line-dwelling* (2018), consisting of 100 parallel 4-metre lengths of thin steel rod, supported on delicate pyramidal props made of cast wax and brass. Each of these drawings was full-scale: they were not depictions of something numerically enlarged or diminished; however, they evoke different scalar imaginations, opening to progressively larger spaces and times, more intricately balanced forces and a higher degree of repetition. *1 line-dwelling* was inhabited by being worn; *10 line-dwelling* implied the possibility of ten wearings in consecutive days or by ten people simultaneously; and *100 line-dwelling* registered the movements of many people, vibrating and sometimes toppling as the floor responded to use.

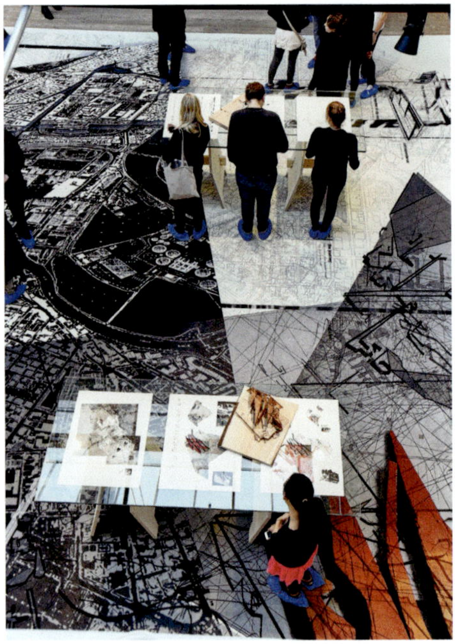

Figure 4.3 *On the Surface*: view from balcony, 2014, Metis: Mark Dorrian + Adrian Hawker. Photograph by Gert Skærlund Andersen.

For *1000 line-dwelling*, the fourth work of the series, Gallagher carefully collected 1,000 silver hairs, fallen from her own head; these were painstakingly knotted together into a single line, and stretched back and forth across the gallery, wound around pin-nails and secured to the walls (see Figures 4.4 and 4.5). Silver hair, microns wide and without the colour lent it by melanin, is almost transparent: these are faint lines. Lengths of fine silver jewellery chain were suspended from this hair line, pulling it downwards (recalling catenary models) (see Figure 4.6). Because a chain is entirely without compressive strength or internal tensions, it hangs precisely vertical; and because the hair lines are so thin as to be invisible, the silver lengths appear to float as if drawn onto the air itself. Sunlight revealed the hair line, so the work registered the passage of the day and conditions of the sky. Gallagher writes:

> The silver chain pulled the slack loops of silver hair into tension, like a builder's hand pulls a string-line soaked in pigment, releasing kinetic energy to form marked lines on a construction site. The taut hair-lines installed at eye level became a generative force, reterritorialising the exhibition site, creating a barely visible cartography of resistance and embellishment.[22]

Matter from Gallagher's own body became the architecture of the space, delineating it, framing and enclosing viewers, supporting structural weight and mediating environmental conditions. It recalls works such as Tomás Saraceno's *Galaxies Forming Along Filaments, like Droplets Along the Strands of a Spider's Web* (2008) in which cellular clusters of black elastic cords connect walls, ceiling and floor; and Gertrude Goldschmidt (Gego)'s (1912–94) *Reticulárea* (1969) with its mesh of aluminium and steel wires which present visually scaleless lines for viewers to navigate. In Saraceno's case, lines evoke the connection between vast and small.

The entire construction had a sense of fragility and lightness; deinstalled, it could be wound up into a single spool and held in the palm of Gallagher's hand. Yet because of this, it has a strong presence. Viewers – line-dwellers – became highly aware of their own bodies and the risk that they might catch on an invisible thread and break the single line, potentially causing the entire delicate construction to collapse. Movement through and within this space was hesitant, slow and deliberate as viewers were induced to see their own bodies as clumsy, heavy and potentially disruptive. Scale, as Gallagher deploys it, is a matter of how we imagine our bodies. Rather than a ratio between large and small, scale manifests between fragile and forceful, singular and multiple. Once the knotted lines are registered as hair, the work is contextualized as recalling the time of human growth and aging and the labour of gathering and knotting. To enter such a drawing, Gallagher suggests, is a form of dwelling: not simply enclosure, but imaginative residence; she writes: 'Lines, animated through force, become drawn ornamental dwellings, and are my imagining of a relational scalar dynamic between cosmos, building and my matter-body.'[23]

Figure 4.4 *1000 line-dwelling*, Sue Gallagher, 2019, silver chain, hair. St Paul St Gallery, 2019. Digital photograph by Sue Gallagher.

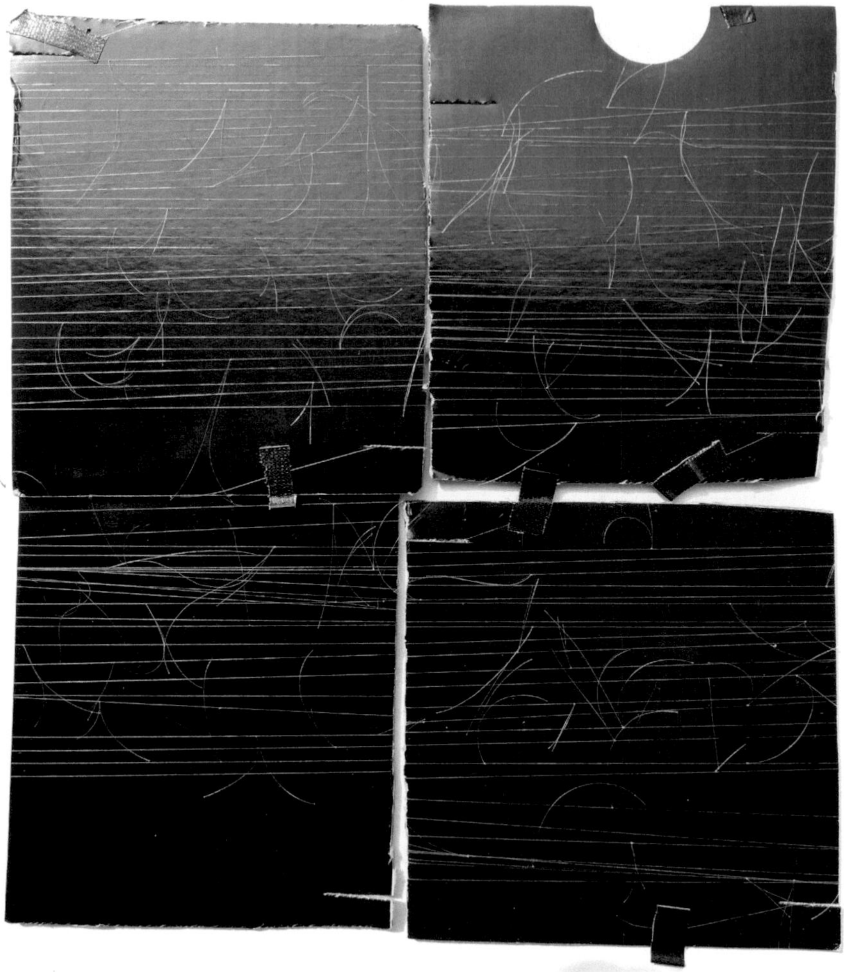

Figure 4.5 Knotted hair-lines for *1000 line-dwelling*, Sue Gallagher, 2019–20. Digital photograph by Sue Gallagher.

For al-Mansur, walking through the plan to simulate the city's spaces was not enough. The invisible order of the proposed city was revealed by standing back and setting fire to the drawing. The dislocated view was also part of his imaginative inhabitation. Architectural drawing always carries with it the potential to multiply views. The simulated, full-size encounters of VR and Lifesize Plans collapse this multiplicity in service of a single coherent vision. They give the sense that a projected work is only multiple or ambiguous temporarily, that it becomes singular and complete on construction. *On the Surface* reveals the vertiginous experience of inhabiting many collaged views at once. Its viewers are provisional figures, temporarily scaled in relation to the fragmentary works they

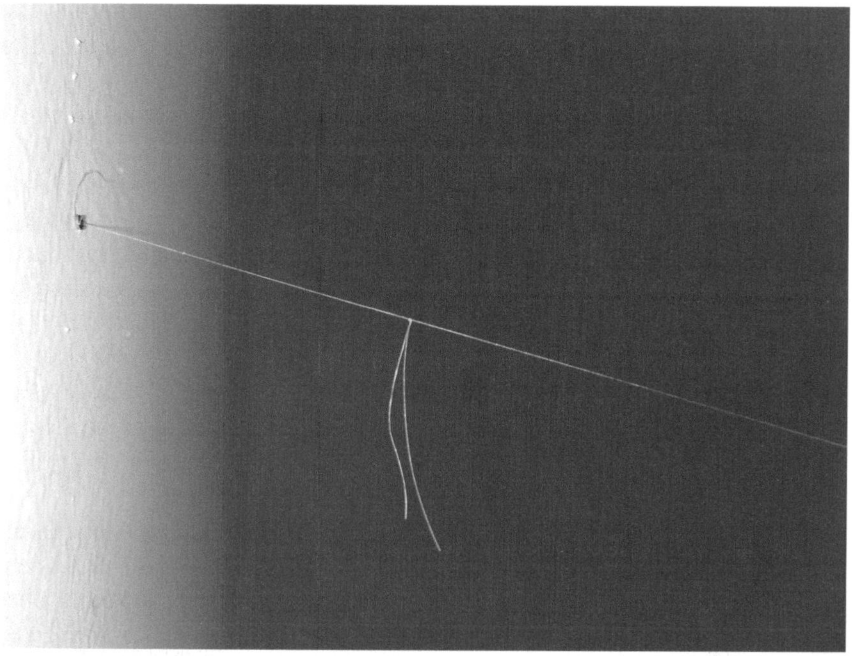

Figure 4.6 *1000 line-dwelling*, Sue Gallagher, 2019, silver chain, hair. St Paul St Gallery, 2019. Digital photograph by Sam Hartnett.

encounter. In *1000 line-dwelling* there is no sense of fragmentation. Instead, the viewer is suspended, moving in slow motion, alerted to delicate shimmers and air currents, drifting in scalar reverie. The drawing's inhabitant is not singular and fixed but suspended in ambiguity.

Hinge and tether

In linking the irreducibly concrete to the ambiguous, the full-scale drawing is like a tether fixed at one end and loose at the other, or a hinge flexing around its pin. The full-scale drawing addresses the specific body of the viewer in a particular place and time and offers a field of encounter rather than a single experience. Inhabitation involves a degree of latitude, scope and possibility. This is more than just the option to view the work from various positions, to choose how to interact or to interpret it differently. Bishop claims one of the crucial effects of installation is 'decentring' of the viewing subject;[24] installation specifically activates the viewer in various ways that may conflict with one's assumed sense of self and situation, producing a sense of dislocation.

Architectural drawings rarely ask their audiences to reflect on their bodily relationship to them, opting instead for a simulated immersion (as if it were real) or an abstracted distancing (via heavily coded notational forms). The drawing-as-hinge, however, does not treat the viewer as a fixed subject but as a flexible one. By refocusing on the observer's interactions with drawings and models, architectural installations frame our engagement with them as sites of encounter.[25] These encounters are not locked in space and time but are at once projective, anticipatory, referential *and* corporeal, immediate, situated. A series of architectural installation-drawings, led by academics Simon Twose and Jules Moloney, specifically address the imaginative inhabitation of spaces that cannot be physically inhabited or directly viewed: the ocean floor, the watery plenum of the ocean and latent seismic topographies.

In *Canyon* (2018), exhibited at Palazzo Bembo concurrently with the 16th Venice Biennale of Architecture, Twose and Moloney, along with Anastasia Globa and Lawrence Harvey, explore a vast underwater space – the Kaikōura Canyon off the coast of Aotearoa New Zealand's South Island.[26] This location achieved prominence due to a violent earthquake in 2016 that fractured the coastline and lifted a huge tranche of the seafloor above the waves, triggering undersea landslides and flows of sediment. Latent seismic activities and ruptures are particularly prominent in the imaginations of those who inhabit such earthquake-prone landscapes, yet these phenomena remain ungraspable. The drawing is a 'projective portal' to inaccessible spatial conditions.[27] The canyon can be imaged through bathymetric surveys but cannot be seen directly, even by an underwater observer. The creators write: 'The canyon landscape is not visible beneath the sea surface, yet its presence is palpable; it is a vast and dynamic material entity, with water kilometres deep, figured by seismic jolts, turbid flows, pressures and intensities.'[28] Drawing places the inaccessible within reach.

The installation, described as a 'multi-modal architectural drawing', consisted of projected imagery, digitally fabricated objects, virtual objects made present through augmented reality and spatialized soundscapes.[29] Spaces were formed and defined by crumpled lengths of tarry black building paper attached to ceiling and walls; these were formed in situ rather than planned in advance (see Figure 4.7). Forming rolls of paper into interior shells was a laborious, awkward process of shaping space from the inside (similar to Kiesler for his *Endless House* in Chapter 3). Pinned to the wall at intervals, the material hung loose or crumpled in between. The interior was not defined by abstracted lines but by the physical behaviour of linear material in response to localized fixings. Gaps opened between the sheets, revealing four small video displays of the canyon as a virtual environment; the same space could also be viewed spatially in VR headsets (Figure 4.7). The various manifestations of the canyon were based on a series of hand-drawings made by Twose. None of the drawings or models of the canyon are literally scaled replications of it. Rather, its makers describe it as a work of three-dimensional 'matter-sketching'.[30] A sketch is, for them, 'a way of exploring, yet not fixing, space and projecting an irresolute architectural sense to domains

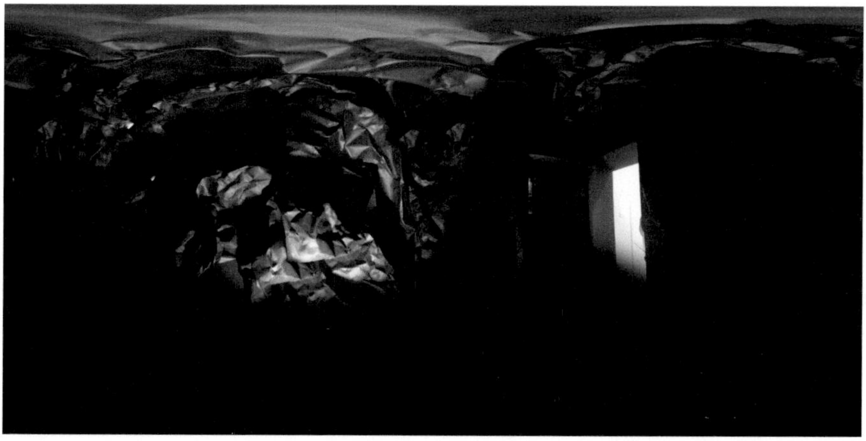

Figure 4.7 *Canyon*: installation view, Palazzo Bembo, Venice Architecture Biennale, 2018. Simon Twose and Jules Moloney with Anastasia Globa and Lawrence Harvey. Photograph by Anastasia Globa.

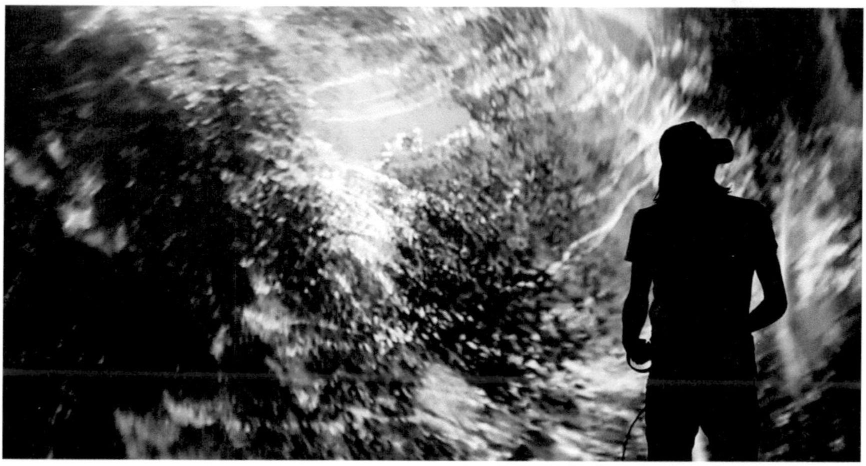

Figure 4.8 *Canyon*: VR sketch environment 2017. Simon Twose and Jules Moloney with Anastasia Globa and Lawrence Harvey. Photograph by Anastasia Globa.

beyond it. The sketch accepts possibilities, blurred approximations, *pentimenti* and mistakes. . . . Irresolute feedback from drawing media is a creative collusion between drawer and other-than-human domains.'[31] Sketches are a form of drawing characterized by looseness and quickness; at the same time, they are valued for their succinctness and efficiency. A sketch is approximate but exact: it 'has zero precision, and consequently zero margin for error'.[32] It is gestural and suggestive, essential and incomplete.

In the case of *Canyon*, sketching enables and intensifies ambiguity, misalignment and approximation. *Canyon* is not a definitive or objective map of the subterranean rift. Rather than endeavouring to fix the canyon in place, sketching holds it in abeyance, *unfixing* it so that it hovers just outside the scene as a shadowy catalyst for experience. The exhibitors write: 'The poiētic power of drawing, as a portal, leads to questions of what qualities are translated from drawing to projected space, and what qualities observed in the space transfer back to drawing . . . drawings have the potential to be open, poiētic portals, projecting not just instrumental understandings, but those at the edge of representational capture.'[33] Drawing here is not a matter of clarification: it is not the revelation of a consistent, underlying order, nor the prediction of how some future body will experience a planned space; it is not problem-solving. Rather, this drawing mode suggests spatial conditions that hover at the edges of awareness, unfixed and intangible, felt rather than known. Immersion in this case is not a persuasive simulation of reality, but the sense of being encompassed by an ungraspable field. The drawing is 'irresolute' not because its makers were insufficiently clear but because irresolve and unclarity are defining features of the space they seek to place us in touch with.[34]

One of the possibilities of installing and exhibiting work is its cyclical redocumentation. *Canyon*, as exhibited in 2018, was recorded using a panoramic camera, and wax casts were made from the crumpled paper surfaces. These were adapted for re-use in ongoing iterations; one such version was experiencing the installation using VR, producing 'the odd effect of allowing one to inhabit the installation with one's previous self'[35] (see Figure 4.8). Just as Mulla, Paterson and Macken employed motion capture to develop lines of continuity between discrete architectural projects (see Chapter 3), *Canyon* acquires a layered history. Twose and Moloney liken this to *pentimenti*, the overlaid lines of a drawing that reveal where a drawer has reconsidered, repented or tried again. Through these remakings, viewers and their experiences become inscribed into the work.

This inscription might even be as a kind of void or gap in the scene. It has often been noted that architectural drawing involves an impossible viewer. Orthographic drawing, in particular, is distinguished from perspective because the viewer is not positioned in space but is infinitely distant. In their embodied architectural drawing *Everest Death Zone* (2013), creative-practice researchers James Craig and Matthew Ozga-Lawn explore this ejection by analogy with a famous fall. In 1924, British mountaineers George Mallory (1886–1924) and Andrew Irvine (1902–24) died during an attempt to reach the summit of Everest. They were last sighted on the summit ridge, when the clouds briefly parted, as two tiny black spots moving high on the mountain.[36] It remains unclear whether they summited and died on the way back, or never reached the top. The death zone is the altitude beyond which a human body cannot survive for long and in which an error or accident will almost certainly be fatal; on Everest, this zone begins at 8,000 metres. Mallory's frozen corpse was finally found seventy-five years later.[37] Craig and Ozga-Lawn exploit this ambiguous relationship between Mallory's body and the summit, and the idea of an uninhabitable space, to question the nature of architectural vision. They write, 'Through the readmittance of the body into consideration . . . we are made more conscious of the positions taken up around and between architectural drawings, of the drawings' limits and their relationships to the body.'[38]

Like *Canyon*, *Everest Death Zone* invites us to consider how inhabiting spaces does not necessarily mean being physically contained by them. Initial drawings mapped the location of Mallory's found body: the mountain was drawn as a fine, triangulated mesh, a geometric net of points rather than a geological mass; weight was drained away and the summit omitted (see Figure 4.9). A photograph was composited into the drawing, looking towards the summit from the point where Mallory's body was found (potentially Mallory's last view). Next, a silhouette of Mallory's body was reconstructed from another photograph. It was the most solid, visual element of the drawing but was hard to map to the form of a body: it seemed to have inhuman angles and thin protrusions. It was separated from the ground and appeared as if floating or in freefall. Mute and unrecognizable, it was almost a landscape itself. Frozen bodies on Everest are difficult to remove and become macabre landmarks. The drawing was structured by Mallory's

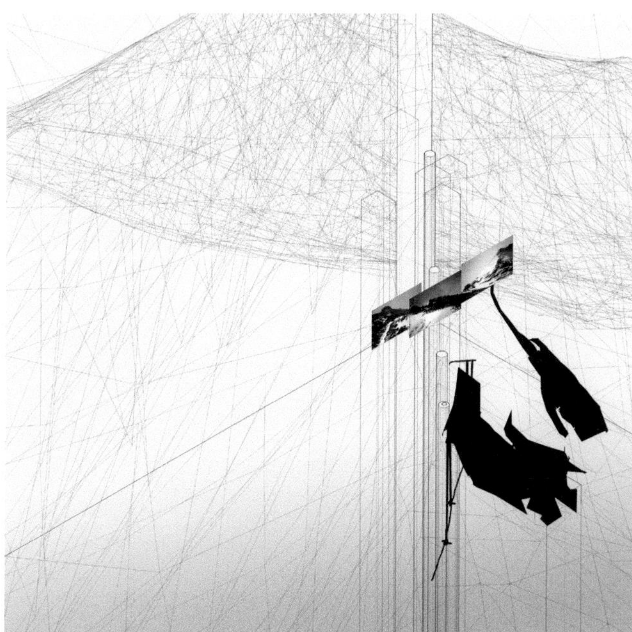

Figure 4.9 Mallory drawing, 2016, Stasus (James A. Craig and Matthew Ozga-Lawn).

intentions. The projected line towards the summit and the objectification of Everest as something to be conquered were reversed by Mallory's fall and death. Architectural drawing is allegorized as a risky, hubristic venture attended with the possibility of failing. *Everest Death Zone* was a kind of *memento mori*. Empty space, it seems to suggest, is not a neutral void waiting to be filled, but a fatally charged expanse in which we struggle for purchase.

Everest Death Zone evolved through reshowings and developments, similar to *Canyon*. In a second phase, the drawing – up to now two-dimensional – was spatialized as a site-specific installation in a large void inside one of the towers of Tyne Bridge in Newcastle-upon-Tyne, UK (see Figure 4.10). The new site was overlaid with qualities of Mount Everest: the height of the space provided similes with altitude, void and the risk of falling. Where gallery spaces typically present themselves as neutral backdrops, this temporary exhibition space was dimly lit, unheated, leaky and shaken by traffic, inviting a more specific relationship to the work. The expansive exteriority of the Himalayas was folded inwards as a dark void. Membranes, reminiscent of tarpaulins and the technical fabrics of mountaineering, were suspended, twisted and tensioned in the interior void using ropes and carabiners. The fine graphic lines of the earlier digital drawings became nylon cords, the drawing medium of mountaineers. (We might recall that cords stretched across the building site were the original architectural lines.) The installation overlays two spaces, across a span of time and distance, hinged or tethered by bodies: Mallory's, the catalyst for the work, and exhibition visitors' bodies.

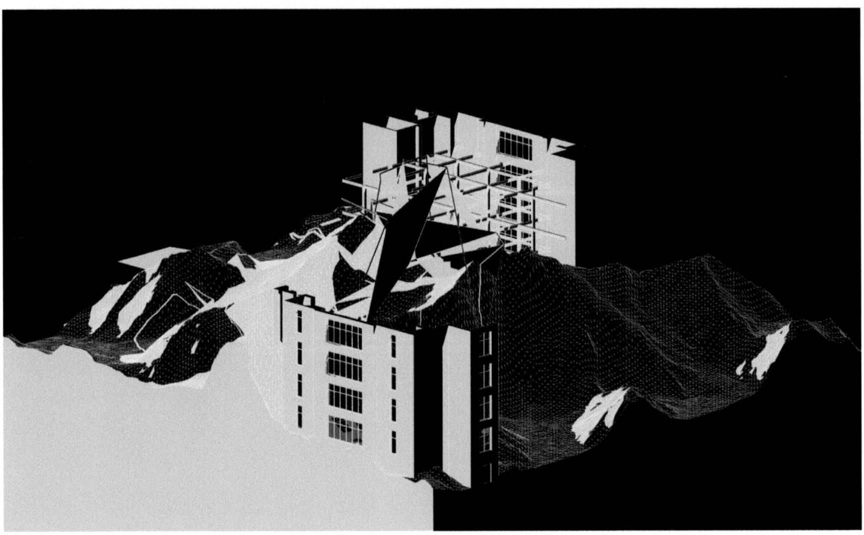

Figure 4.10 3D model merging Everest landscape with Tyne Bridge North Tower, 2016, Stasus (James A. Craig and Matthew Ozga-Lawn).

Clips from a 1924 documentary of the expedition were projected on the twisted planes that resulted.[39] Critically, the viewer struggles to find somewhere to stand to view the image. Usually, film projections are arranged to provide a clear view from a convenient location. Here, projected into the void, they address the viewer obliquely. The viewer is like Mallory at the bottom of the slope, looking up at a landscape that has escaped him. The predictable perspective constructions of Albrecht Dürer (discussed in Chapter 2) use cords to model the passage of light. Craig and Ozga-Lawn's construction is viewed obliquely, untethering from the viewer's eye. Referring to one of Dürer's engravings (see Figure 2.3), Craig and Ozga-Lawn conclude:

> Here, the artist is represented as being unaligned with his own subjectivity. He is not able to depict the reality of what he sees: the lute, the string, the figure dictating the outline of the drawing, all subjective reality and the fluidity of life is lost due to the power of perspective and the tools required to support this method. What becomes apparent in this removal of subjectivity is that one must destroy oneself in the pursuit of the ideal external object.[40]

Themes of incompleteness, failure, stasis and fall recur in the third iteration of the work: the rope and screen construction were reassembled at a smaller scale in a new site and augmented with VR imagery. In this new iteration, the ropes previously used to construct the suspended screens were used to restrain and tether viewers in physical space while they wore a headset and entered the virtual scene. This space included elements from the topological model of Everest, layered with a model of the Tyne Bridge tower. The two remote and previously unconnected sites continued their journey towards one another, entering into a dialogue that revealed unforeseen alliances. In each iteration of *Everest Death Zone*, the viewer was aligned with Mallory, but, as the exhibitors write, 'as soon as we step away and navigate around the installation, we see the condition of perspective and its machinations exposed'.[41] If perspective is the art of situating the viewer, installation is something more unsettled. Perspective centralizes the viewer, constructing space around them and revealing things to their sight. Installation, however, opens a field of possible encounters: many viewings, uses, participations and enactments. It casts us into more than one space at once.

Half light

Installation also manipulates our relationship to time, revealing multiple, even contradictory temporalities. While clock time is abstract and technical, experiential time is corporeal and social, and embedded in materials, processes and practices. Architectural installation-drawings provide an opportunity to access these times which are typically invisible in conventional drawings.

In *AirXY* (2008), for the 11th Venice Biennale of Architecture, Erik Adigard and Chris Salter (M-A-D) imagine the ubiquitous surface of the screen as a habitable space-time. Screens have proliferated, becoming central to the way we understand images. Where time might once have been implied through image sequences, repetitions or marks of age, screens change before our eyes, often responding immediately to our own actions. As we suggest in Chapter 2, the screen enables the possibility of interfacing directly with images. Adigard and Salter imagine that we might physically enter into the screen. Viewers were greeted by a wall-sized screen surface, operating as a clock with four layers. On the first was a series of screenshots from desktop interfaces; over these was superimposed a clock in the form of an XY grid of 24-hour and 60-minute units; in front of these was a layer of randomly floating icons; and finally there was a surveillance layer: visitors, sensed and monitored by overhead cameras, were projected in real time as moving forms[42] (see Figure 4.11). Behind the screen was a dark, hazy space with these moving images projected on the floor. Every minute, a strobe light flashed: turning the space bright white for a split second, creating an after-image, which then re-materialized.[43] For a moment the viewer was made uncertain about whether the image was on the screen or merely a ghost on their retinas. Smoke-machines thickened the air so the projected imagery could be seen in space. The screen – in prosaic terms a wall

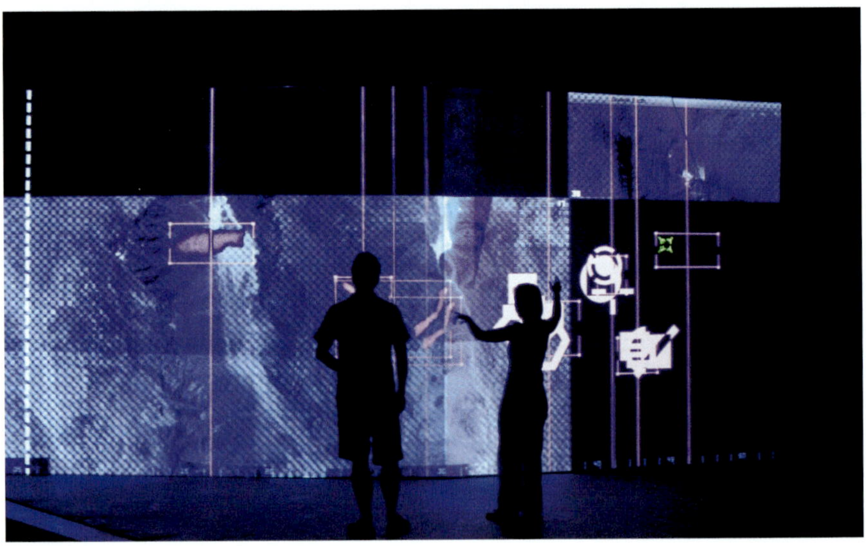

Figure 4.11 *AirXY* diptych, 2008, M-A-D, Erik Adigard and Chris Salter. Photograph by Anke Burger.

in the gallery – was dematerialized or rendered ambiguous. Adigard and Salter intended the surface to dissipate into 'a responsive, inhabitable Ether'.[44] The work gave spatial effect to the intuition that we live through our screens. Time was occupied as a space. *AirXY* was perhaps a pre-vision of a future possibility in which the screen becomes atmospheric, or perhaps a critical revelation of the frenetic, immersive quality of contemporary environments.

Full-scale drawing acts in the present tense. In *AirXY*, flickering and flashing set a stuttering and ceaseless rhythm that always seemed to be moving too fast to keep up with; it washed over the viewer. We can also experience time as immersive by slowing the present, as another work by Mulla, Paterson and Macken demonstrates. In a darkened gallery in Ōtautahi Christchurch, Aotearoa New Zealand, a 12-metre-long, bead-blasted aluminium truss frame slowly moved back and forth across the floor, driven by a small concealed electric motor.[45] The frame tapered to a point tethered to the gallery wall; its larger end was mounted on rubber tyres (see Figure 4.12,4.3). An algorithmically generated soundscape accentuated the structure's movement through the space. Varying in volume, timbre and pattern, it recalled droplets of water, chimes or audio glitches. The work – *Drawing Machine* – resembled a compass patiently describing an arc across the floor; its tractor tyres also suggested an agricultural irrigator. As half a vehicle, it was mobile at one end but held at the other, tracing its line again and again, leaving a rubbery track on the gallery floor that resembled the conventional door-swing indication on an architectural plan.

Viewers moved more quickly than the drawing arm, so it seemed to be creeping slowly. Changing patterns were revealed as viewers lingered. In its eight-minute cycle, *Drawing Machine* passed through beams of light; the frame's structure, and aluminium scoops within it, filtered this light (see Figures 4.12; 4.13). When in shadow, the matte metallic surfaces tended to disappear in the dimmed gallery. When a body passed in front of this projected shadow-drawing, it caught the light differently, producing new spatial lines. There has been a long understanding of the close relationship between shadow-lines and drawn lines due to the Greek legend of drawing's origin of Diboutades tracing her lover's shadow-profile on a wall.[46] The geometry of single-point perspective, in which lines are stretched out towards a vanishing point, correspond with rays of light radiating from a point-source. Accurately drawing shadows (skiagraphy) requires constructing the intersection of these two geometric fans. If the truss is a drawing machine, the shadows are the drawing and the viewers' own shadows inhabit it.

The underlying proposition of *Drawing Machine* is that architecture might be drawn in light. In their built works, Mulla and Paterson often employ shadows to flatten three-dimensional objects onto surfaces to emphasize profile and tone. These changing conditions are ephemeral and cyclic: 'In passage alcoves you stand in darkness looking through light coming from the window, toward a plane of darkness. The effect is a stratum of interior darkness with rich views towards

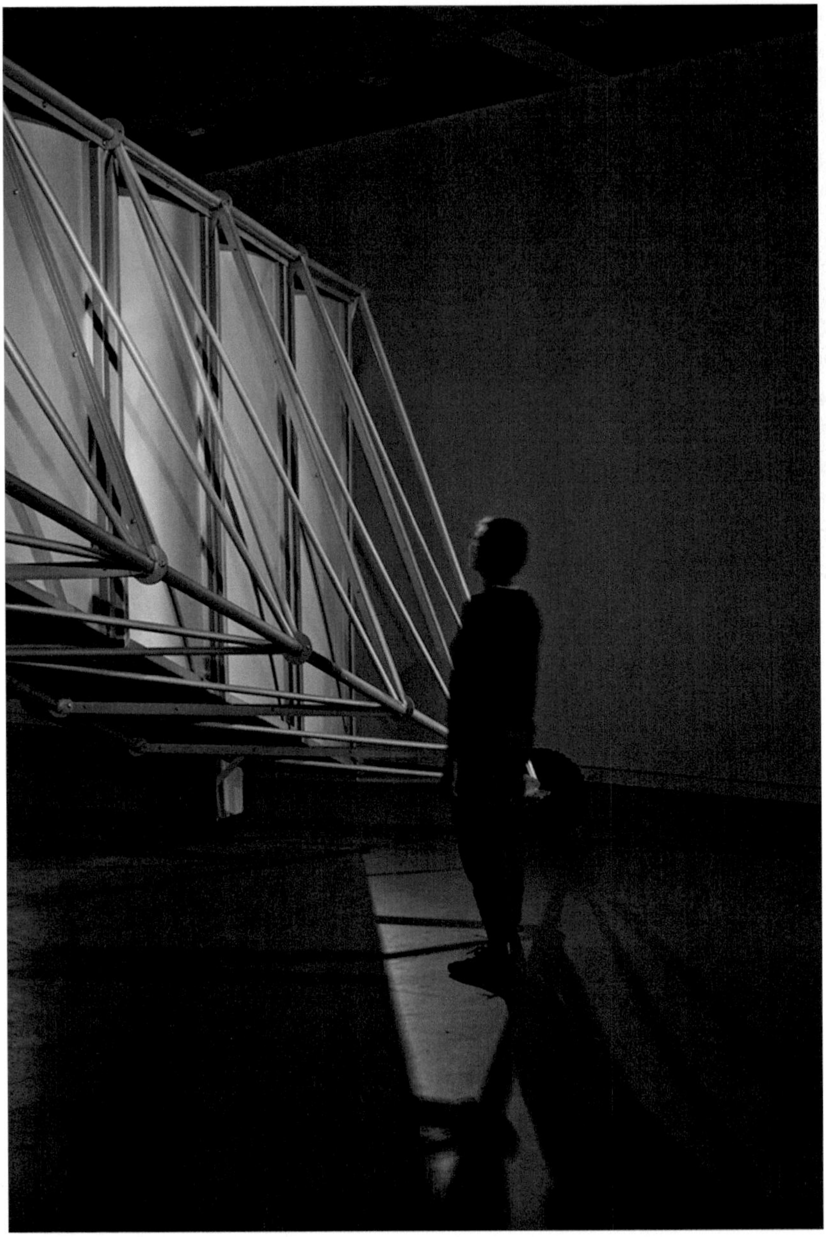

Figure 4.12 *Drawing Room*: detail of *Drawing Machine*, Toi Moroki Centre of Contemporary Art, Ōtautahi Christchurch, Aotearoa New Zealand, 2020, Sarosh Mulla, Aaron Paterson and Marian Macken. Bead-blasted aluminium frame, rubber tyres, electric motor. 39 ft. 4 7/16 in. × 10 ft. 5 63/64 in. (12 × 3.2 m). Photography by Simon Devitt. © Simon Devitt Photographer/Pac Studio.

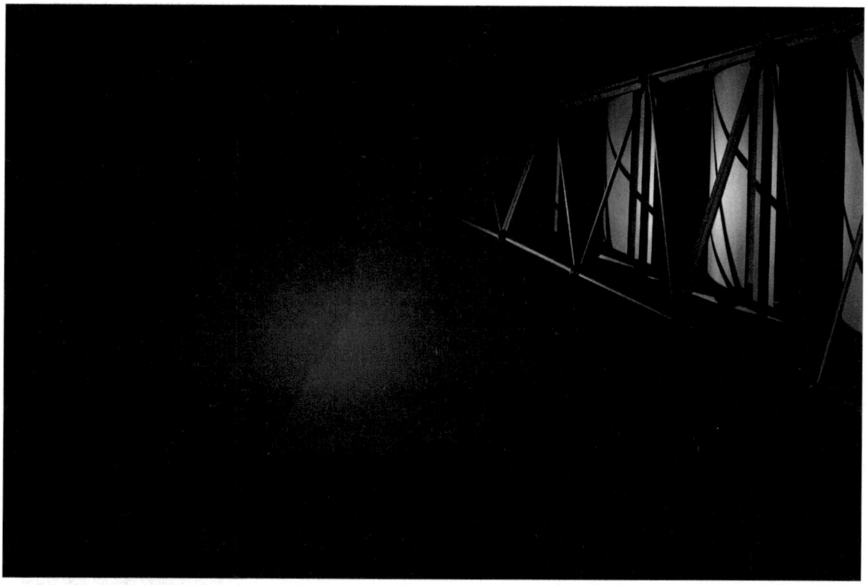

Figure 4.13 *Drawing Room*: detail of *Drawing Machine*, Toi Moroki Centre of Contemporary Art, Ōtautahi Christchurch, Aotearoa New Zealand, 2020, Sarosh Mulla, Aaron Paterson and Marian Macken. Bead-blasted aluminium frame, rubber tyres, electric motor. 39 ft. 4 7/16 in. × 10 ft. 5 63/64 in. (12 × 3.2 m). Photography by Simon Devitt. © Simon Devitt Photographer/Pac Studio.

the light. . . . The shadows of architectural components fall on the house's exterior and surrounding ground, and the adjacent foliage brings shadows to the interior.'[47] The house is a constantly operating drawing machine. In their catalogue essay, they refer to architect Lynda Simmons's view that shadow and interior intensity are particularly important architectural conditions in New Zealand and the Pacific. Simmons describes how the darkened interior of the Māori wharenui (meeting house) is approached across a bright open expanse known as marae atea. The latter is a place of exterior relation, meeting and debate. The interior of the wharenui is a place of introspection and genealogical connection, likened to the interior of a body. Shadow 'provides a quality of fullness', while 'glare, separating and connecting the inner realm and the illuminated plane beyond' demarcates thresholds.[48] In a similar way, Pac Studio draw with light in darkness, partially revealing things hidden there. Drawing with shadow in light suppresses form, revealing by subtraction.

Fascination with the 'interior intensity' of being immersed in shadow was also shown in an earlier work by Mulla and Paterson, titled *Penumbral Reflections* (2018).[49] Here, a cubic steel frame sat in the centre of a square gallery, centrally illuminated to cast linear shadows on the walls (see Figure 4.14). These mingled

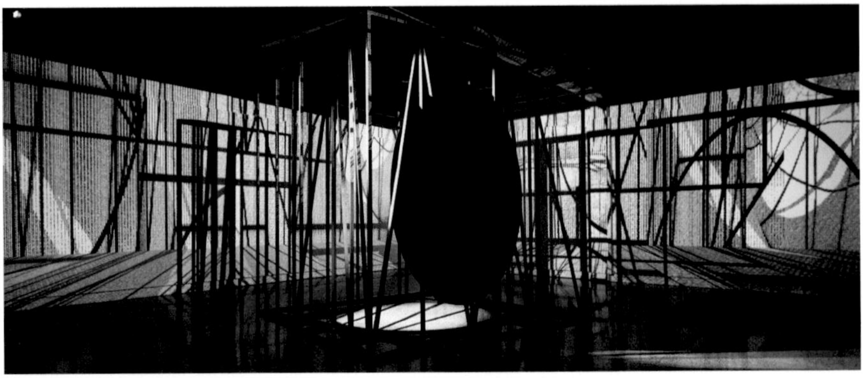

Figure 4.14 *Penumbral Reflections*, Objectspace, Tāmaki Makaurau Auckland, Aotearoa New Zealand, 2018, Sarosh Mulla, Aaron Paterson and Pac Studio. Photograph by David St George.

with projected shadows rendered and animated using a digital model. This created ambiguity; it was hard to tell at a glance which shadows resulted directly from light interacting with physical matter and which from a simulated performance of light. The viewers' own shadows were added to this panoramic scene. Presiding at the centre of this optical field, held within the frame, was a 4-metre-tall black convex mirror. Fabricated from stainless steel, using a technique for shaping the endcaps of milk-tanker vessels, and lacquered, the mirror collected light, capturing the scene in miniature. It also focused sound, creating invisible acoustic hotspots. *Penumbral Reflections*, like *Drawing Machine*, was revealed by the moving presence of the viewer.

We see, in these installations, versions of traditional architectural equipment: the screen, a mainstay of contemporary drawing, and the more-or-less obsolete drawing machine. In Chapter 2, we deliberately muddied the distinction between a *drawing* and a *drawing apparatus*; we argue that Dürer's temporary constructions of cord, frames, styluses and hooks; Alberti's *goniometer* and *finitorum*; and the surveyor's ropes and stakes could be understood not just as equipment used to make a drawing but already the enactment of a full-scale spatial drawing. Both *AirXY* and *Drawing Machine* hold out the possibility of becoming bodily immersed in the time of drawing, suspended in the present tense.

Habitable drawings

Fred Sandback (1943–2003) described his spatial cord installations (see Chapter 1) as 'a diffuse interface between myself, my environment, and others peopling that environment, built of thin lines that left enough room to move through and around. It was still sculpture, though less dense, with an ambivalence between exterior and interior. A drawing that is habitable.'[50] He saw his habitable drawings as interfaces mediating between bodies in a specific situation. As interfaces, they created connections or passages. Crucially, they left room for movement; they held a space open but did not fill it. The architectural installation-drawings discussed in this chapter are similar: they mediate, connect and are held open, fostering a degree of ambiguity. This ambiguity and openness, we suggest, is fundamental to inhabitation. To inhabit is not merely to be enclosed or immersed, but to be situated or contextualized. Such an understanding is not new to architecture; Juhani Pallasmaa writes of our encounter with a building: 'it is approached, confronted, related to one's body, moved through, utilized as a condition for other things. A building is not an end in itself; it frames, articulates, gives significance, relates separates and unites, facilitates and prohibits.'[51] The building is not just experienced as a sum of perceptions but in terms of the possibilities it affords us, the context it provides and the way it offers its own context to us. Merleau-Ponty summarizes: 'Thus, we must not say that our body is *in* space, nor for that matter *in* time. It *inhabits* space and time.'[52]

Conventional VR visualizations and Lifesize Plans' projections try to dispel the present place and time in favour of a simulated future, as if the viewer could be transported there. VR offers a simulated bodily experience, through a digital twin, that emphasizes the visual and perhaps auditory and proprioceptive senses. Strategies, such as employed by Lifesize Plans, trade off visual fidelity to emphasize how distance and physical orientation feels. The other works considered in this chapter, by contrast, point elsewhere, but are *also* situated enactments. They are full-scale and present tense, in that they address the viewer's physical presence at the moment of encounter; but this does not mean that they offer a single space or time. On the contrary, manipulation and layering of space and time are their hallmarks. Nor do they address the body as a singular presence: it is not *only* a sum of perceptual experiences, nor simply an organism to be accommodated. Elizabeth Grosz claims incompleteness and indeterminacy as defining aspects of the body. The body integrates experience but it also resists integration:

> The body is a most peculiar 'thing,' for it is never quite reducible to being merely a thing; nor does it ever quite manage to rise above the status of thing. Thus it is both a thing and a nonthing, an object, but an object which somehow contains or coexists with an interiority, an object able to take itself and others as subjects, a unique kind of object not reducible to other objects. . . . Bodies are not inert; they function interactively and productively. They act and react. They generate what is new, surprising, unpredictable.[53]

The term *viewer*, as we indicated at the start of this chapter, is a placeholder term intended to capture the multifarious ways a drawing might be received or encountered. A more accurate term might be *inhabitant*. In comparison to the commonplace architectural scale figure or anthropometric body, actual bodies are unpredictable, not normative. There is always another possible permutation of body, another set of physical capacities and constraints, another social, psychological, cultural or ecological construction of body that could come into play. Window, portal, interface and screen remain persistent metaphors for architectural drawing. Each implies a degree of transparency – we look through the drawing to another place. But each also implies distance, displacement or separation. The inhabitable drawing acts back through the window, inviting us to reconsider our own position and situation.

5
DRAWING ROOM
EMBODIMENT IN STUDIO

Studio as situated inquiry

One of our motivations for writing this book is that we are teachers. In this chapter, we look at full-scale drawing as a way to learn in architecture and related fields and reflect on some of our own experiences of full-scale drawing with students. Speculative drawing is sometimes criticized for its abstraction and disengagement from the professional practice of architecture. To the contrary, we take the position that full-scale experimental drawing enables students to engage more directly with, and develop richer understandings of, architecture and other forms of space-making. Pedagogically, we see it as emphasizing a bodily sense of architecture and space as opposed to a predominantly visual or schematic one. We believe these expanded drawing practices develop an understanding of *situation*: students learn to critically question and understand spaces they directly encounter. Crucially, we see full-scale drawing as orienting students towards creative *inquiry*. We teach in Aotearoa New Zealand: at Waipapa Taumata Rau University of Auckland's Te Pare School of Architecture and Planning and Te Wānanga Aronui o Tāmaki Makau Rau Auckland University of Technology's Te Kura Toi a Hoahoa School of Art and Design. Our examples here are intended to show some of the pedagogical possibilities of full-scale architectural drawing, but we do not assume these are universally applicable. We share them in the hope they might prompt others to develop their own strategies for teaching and learning in the field of architectural drawing.

Learning to design is not simply a matter of acquiring predetermined sets of knowledge and skills. In order to design for as-yet-unknown situations, a designer needs to develop a flexible repertoire of strategies for improvisation, discussion, evaluation, motivation, self-critique and creative discovery. Studio-based teaching typically involves the setting of a design problem or focus on a particular situation to which students apply their own investigation, response or solution. It is sometimes assumed that the more concretely and specifically this problem is defined, the more applicable learning will be to designing in the 'real world' (a place believed to exist elsewhere from the university, characterized by its practical constraints and commercial imperatives). However, deliberately vague, conflicted and imaginative design situations allow students to form their own protocols for interpreting scenarios, identifying what it is that they need to learn and from whom to produce constructive contributions. As Stephen Temple writes: 'Learning design is in the doing, a process in which direct engagement and active involvement in learning for yourself will enable construction of your own ideas about what you are doing as you do it.'[1] Research and design are themselves situated and embodied activities.

The teaching design studio, as both a place and a set of associated practices, already explicitly includes the body.[2] Much studio learning is tacit and embodied, a matter of becoming sensitive to one's surroundings and process and developing intuition. As places of communal learning, studios hold discussions and bodily encounters between teachers, students and the evolving work.[3] They are performative, hosting a series of engaged acts between educators and students with bodies as the principal medium for creative encounters with architectural work. In his classic study of the design studio, Donald Schön (1930–97) writes of 'a reflective "conversation with the situation"'.[4] The various things made and done in the studio – drawings, models, renders, diagrams, samples, texts, photographs, etc. – respond as they are re-encountered in this shared space. The premise of studio-based teaching is that design is 'an embodied language of learning and practising that promotes an ethos of thinking, making, drawing, and exhibiting together'.[5]

Studio teaching is threatened by tertiary institutions' trend towards increasingly corporate-style models of operation. Dedicated studio spaces, often with allocated individual student workspaces, and requirements for specialized equipment and technology, can appear as inefficient uses of real estate and resources compared to courses that can be taught in generic and bookable classrooms. Since the COVID-19 lockdowns of the early 2020s, universities have invested in hybrid modes of course delivery – either as online only or a mixture of in-person with pre-recorded or digitally available content – as a way to increase the pool of potential students and reduce staff and overhead costs of physical classes. However, widely used video-conferencing platforms prioritize a fractured, frontal gaze and interrupt the nuanced bodily aspects of communication. They demand continuous attention diminishing the benefits of silence and gaps in activity.[6] Common university learning management systems (such as Canvas and Blackboard) present learning as a pre-structured ascent towards pre-established aims, organized around modules, tasks and quizzes. This bureaucratic model of teaching – whether online or in-person – reinforces the idea that learning is the completion of tasks. Uncertainty, chance, complexity and failure are taken to reflect the teacher's lack of skill or obtuseness. A focus on completing tasks can be at odds with the fact that some kinds of learning require sustained, ongoing practice. There are links between the pedagogical deprioritization of the body, the atrophy of dedicated studio spaces and the dislocation of the body in architectural representation. As James A. Craig, Christos P. Kakalis and Matthew Ozga-Lawn put it, the 'disappearing act of the body is pertinent to discourses on architectural representation; primarily, the legacy of linear perspective and its capacity to dislocate the subject from the site of a drawing's production'.[7] If design becomes an incorporeal activity, then learning to design is freed of bodily investments.

Design-build projects are sometimes included in courses of study and offer students a chance to have experiences of full-scale built outcomes. However, such 'live' projects present challenges: semester-long university courses may struggle to accommodate the longer timeframes needed by these scenarios; group work and division of roles can be difficult to square with assessing individual achievement; and they can be hard to fund and resource. Long-running instances of design-build studios, such as Auburn University's Rural Studio, are the exception rather than the rule. In the 1970s, student building projects flourished in architectural schools. For example, the Ecological House (1972–5) was built by students from the Architectural Association in London (partly funded by the institution as student-led research); and in the southern hemisphere, University of Sydney students built and lived in the Autonomous House (1974–8) for academic credit[8] (see Figure 5.1). These projects sat in the context of growing interest in collective living, intentional communities and self-sufficiency. They were a manifestation of a global 'anti-school' movement in which student strikes reacted against a range of issues including politics and educational frameworks and resulted in pedagogical experiments.[9] Student-initiated congresses allowed these interests to develop outside institutional influences. For example, in 1972 a student congress was held in the abandoned Mahurangi cement works north of Tāmaki Makaurau Auckland. Modelled on the multi-day music festival, participants gathered in an ad hoc encampment; architecture students built a conference venue, including a cardboard dome, and other structures.[10] The Auckland School of Architecture was described at that time as 'pretty loose' and had devolved into three sub-schools representing distinct polemical positions on architecture.[11] Each centred on a studio space that was modified by students as a shared collective project; the 1972 student congress was an extension of this ethos of self-built pedagogical spaces. In Italy, the avant-garde Global Tools workshops (1972–6) envisioned a radical transdisciplinary

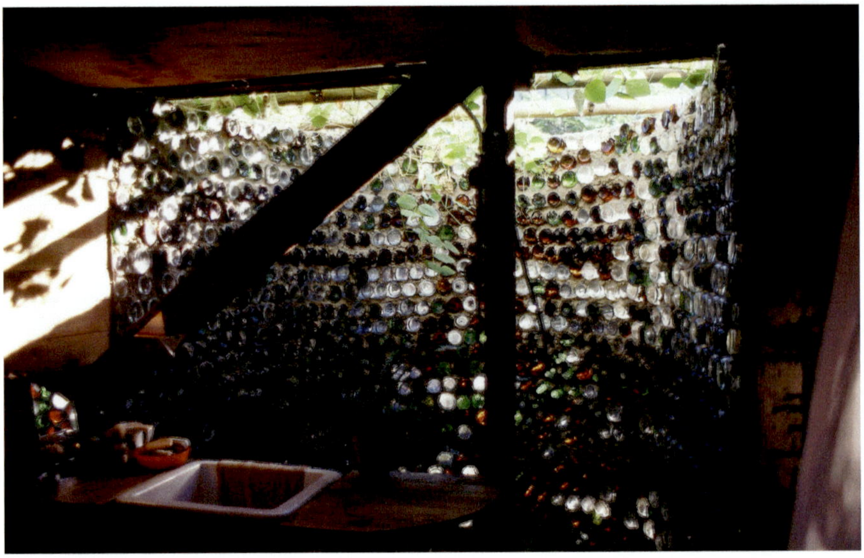

Figure 5.1 Sydney Autonomous House: interior, *c.* 1977. Photograph by Tone Wheeler, courtesy of the School of Architecture, Design and Planning, University of Sydney.

education oriented by artistic investigations.[12] The body was framed variously as the primary instrument for acquiring knowledge, as a language, a tool and means through which to understand situations and contexts.[13] Corporeal activities aimed to liberate the participants from 'physical and intellectual exploitation' and 'cognitive capitalism'.[14]

Such endeavours can seem fantastic compared to the current, increasingly risk-averse, university climate of the 2020s, but recent examples exist. German art and architecture collective raumlabor undertakes a range of self-initiated projects including temporary activations, theatres, fairs and the 'fictional' Open Raumlabor University. Their *Floating University Berlin* (2018), positioned at a rainwater retention basin of the former Tempelhof Airport, became a 'visionary inner city offshore-laboratory for collective, experimental learning', hosting a programme for interdisciplinary learning.[15] Design-build projects can help students develop practical industry skills like building, digital fabrication, project management and professional communication.[16] However, there are also important benefits in full-scale projects that do not directly participate in or simulate the professional world. Full-scale experimentation need not reinforce existing disciplinary structures but can call them into question. Historian Lee Stickells, writing about student-initiated builds like the Sydney Autonomous House, noted that such projects offer 'not finished architectural objects, but a constructive and flexible ground for experiments in being human and being social while also testing ecological boundary conditions. . . . They opened a view to a mode of messy, flawed but ambitious architectural experimentation – exhibiting a sort of improvised, outlaw laboratory praxis.'[17] Full-scale drawing-installations, we suggest, activate this experimentation and testing of boundaries while retaining something of the lightness and provisionality possible in drawing. The following instances from our own teaching practices explore full-scale drawing that reflects critically on the nature of body, embodied learning and the situated inquiry of the studio.

Ungrounding students

The early years of undergraduate studios in design degrees often introduce students to essential disciplinary knowledge and practices of their chosen field, known as foundational or core learning. This involves acquiring common skills, processes, language and norms and reflecting on typical examples. It is our experience, however, that at the beginning of their study, students often need first to be ungrounded or pulled away from assumptions and inherited ideas. An overemphasis on the common, typical or normative may reinforce these assumptions, giving students an unrealistically narrow and fixed view of their field. For students to develop a rich understanding of what is at stake in the design territory, what questions to ask and what might be possible, the field must first be opened to them. Making and acting directly at full scale can be invaluable for this by localizing and grounding the field through and in the student's own body.

In a two-week exercise in a first-year architectural design studio, students were asked to design a wearable space for conversation.[18] The brief for 'Wear: trio/triform/triplet >> wearing a conversation' instructed:

> For this exercise you are both the designer and the inhabitant. You are to design and make an apparatus to be worn by three people. This apparatus will create a space for conversation between three people, almost like a room. Think of it as a 1:1 scale model. The apparatus must: be able to be worn standing up and/or sitting down; and create a distance between each person that is no less than 200mm and no more than 2500mm.[19]

The key driver and primary focus for the task was for students to recognize the body as the location of spatial experience and knowledge and architecture as a practice of mediating relationships between bodies. The *site*, rather than a fixed location, was the body (more specifically, a variable configuration of three conversational bodies). Precedents shown to students included: Allan Wexler's

Four People Wearing a Table (1996), to demonstrate individuals coming together to make a whole; Vito Acconci's (1940–2017) *Instant House* (1980), for its bodily manipulation from flat to horizontal; GAIA's *Veasyble* (2009), as wearable folded paper coverings both joining and isolating; and Franz Erhard Walther's bodily-dependent textile sculptures, such as *Action Pieces* (1968). Rather than projecting the space of future buildings, students were asked to manifest a corporeal and social space directly in the present. Space was, from the outset, not merely a measurable volume but also a place of social and cultural values (and tensions); Karen A. Franke and R. Bianca Lepori write in their book *Architecture from the Inside Out*:

> Even without the presence of a physical boundary, the places we occupy with our bodies, particularly those that we imbue with memories and dreams, become another kind of inside, the *here* of human occupancy and aspiration. Even as we occupy a public place temporarily, possibly a seat on a bus or a train, the space immediately around us, our 'personal space', is a kind of inside, an inside we carry with us, bounded invisibly but symbolically.[20]

Students grappled with scale as a relationship between a single individual, a group of three and everyone else outside this relationship. They mediated intimacy, inclusion and openness. The resulting apparatuses negotiated social spaces between bodies, employed filters and screens and positioned bodies in relation to others using strategies to achieve precision, movement and adjustment (see Figures 5.2–5.4).

The exercise was deliberately fast paced, so students had to work quickly without too much to-scale planning. They constructed their conversational apparatuses and brought them to a picnic in a semi-public campus courtyard. Organized and experienced as an event, the work was positioned in time, and performative aspects, like transporting, setting-up and demonstrating, were given importance. The event itself became a drawing workshop: students drew their apparatuses, making operational diagrams, observational drawings and delineations of the space formed between the three conversationalists (see Figure 5.5). Back in the studio, the latter was converted into a volumetric drawing, a closed boxboard form and then a void in a cast plaster model (see Figure 5.6). With its outer formwork removed, the model became sectional, exposing a translation of the space between the conversing participants to the void of the model. Only at this point were students introduced to the metric logic of scale. The full-scale bodily apparatus was abstracted as a scaleless model, to which metric scale could then be applied. Rather than positioning architecture as something that belongs to an imagined future (which may seem

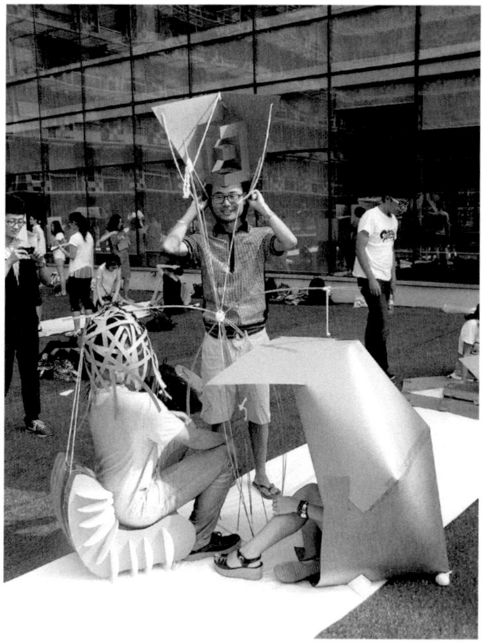

Figure 5.2 Wear: trio/triform/triplet >> wearing a conversation design exercise, Xi'an Jiaotong-Liverpool University (XJTLU), 2013. Photograph by Anna Andrianova.

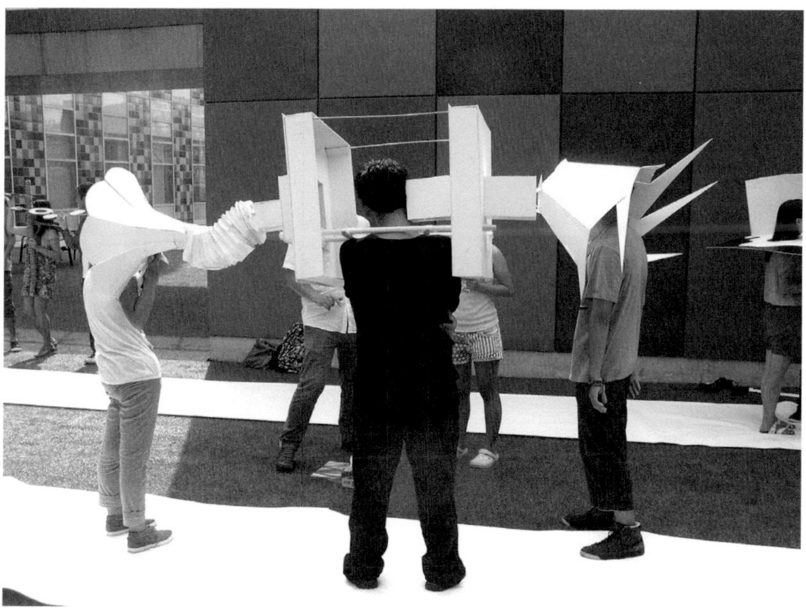

Figure 5.3 Wear: trio/triform/triplet >> wearing a conversation design exercise, XJTLU, 2013. Photograph by Anna Andrianova.

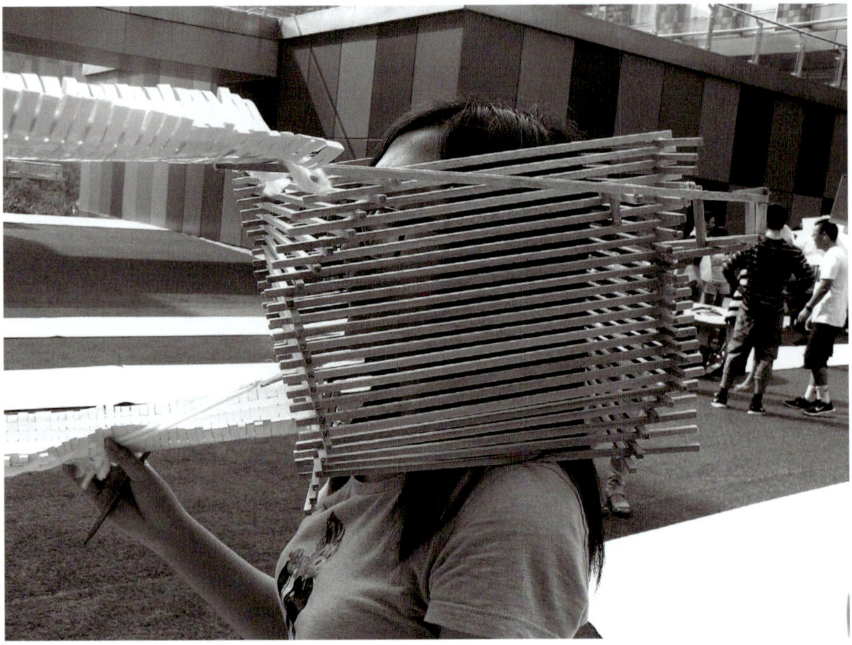

Figure 5.4 Wear: trio/triform/triplet >> wearing a conversation design exercise, XJTLU, 2013. Photograph by Anna Andrianova.

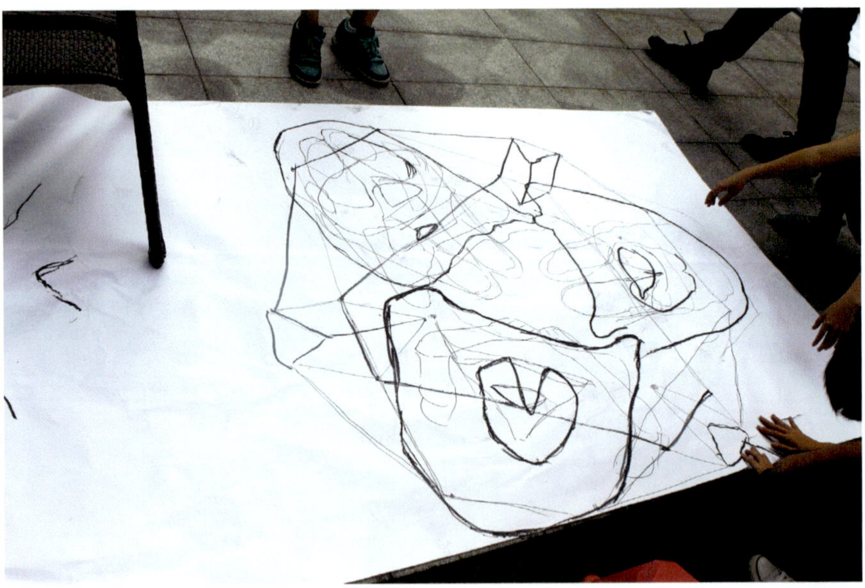

Figure 5.5 Wear: trio/triform/triplet >> wearing a conversation design exercise: drawing exercise, XJTLU, 2013. Photograph by Marian Macken.

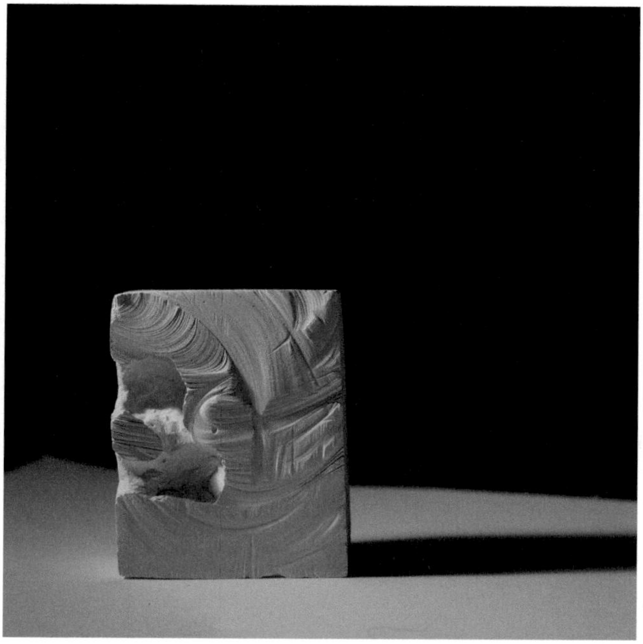

Figure 5.6 Wear: trio/triform/triplet >> wearing a conversation design exercise: plaster model, XJTLU, 2014, Yixin Han. Photograph by Yixin Han.

remote and hard to commit to or invest in), space is first explored as something immediately present but provisional. Students were encouraged to see their apparatus and each subsequent stage of the exercise as a model rather than something final. This iterative approach emphasized time and temporary and changing relationships rather than fundamental concepts like grounding or fixing. Drawings were experienced first, not as plans for the future but as a link in a chain of actions.

A subsequently developed core course for first-year architecture students emphasizes thinking through drawing.[21] While studio-based, this course runs parallel to a core studio project, supporting skills to develop architectural media as a means to catalyse architectural thinking within design strategies. At the beginning of the course, students predominantly understand architectural media as communication for concepts or proposals to others, and then the course introduces students to media as a process of creative and strategic thought. As Paul Emmons and Carolina Dayer write, this requires shifting from the descriptive to the performative: 'The question of performance in architectural drawing is especially critical because design thinking takes place within the practice of drawing. Since architects make drawings rather than buildings, drawing is the central artisanal craft-knowledge of architecture. In addition to being pragmatic, drawing practices are cultural, historical, and even ritualized. They are ways of thinking embedded in ways of operating.'[22]

The course is structured around scale (in both the metric and experiential forms); it starts and ends at full scale. In the first task students measure components of a room with their bodies, using their own particular dimensions as a module (see Figures 5.7 and 5.8). Through elevational photography and digital collage, the body is multiplied to visualize this personal measure. Later in the semester, students arrive at 1:20 scale sections of proposed spatial situations. The following week, they redraw these details at full scale on large sheets of brown paper, a physically involved process (see Figure 5.9); they are then asked to place themselves in the drawing. While some do this through digital montage, the more successful find inventive methods such as spatial recreations, sequential photography, projections of drawings onto their bodies or silhouetted light projections (see Figures 5.10–5.12). In these drawings, full scale is used not to finalize a design but to speculate upon it.

The goal for these activities is to guide students away from normative applications of the body in architectural media. Bodies are too often represented as static, universal, normative and abstract. While a generic body may provide a visual gauge for size or indicate function, it omits forms of inhabitation, disregards bodily diversity and neglects the complexity of corporeality. Through full-scale drawing, a more meaningful, inquisitive and intuitive approach to human bodies is sought. Students come to understand their body as a means to measure, interpret and inhabit. This full-scale drawing practice also fosters collaborative relationships: large drawings are difficult to make alone.

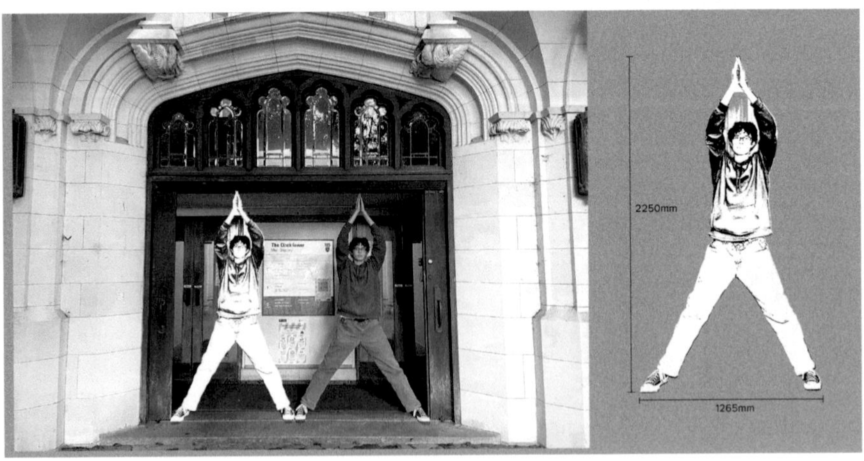

Figure 5.7 Architectural Media 2: Task 1, 2021, Matthew Pritchard.

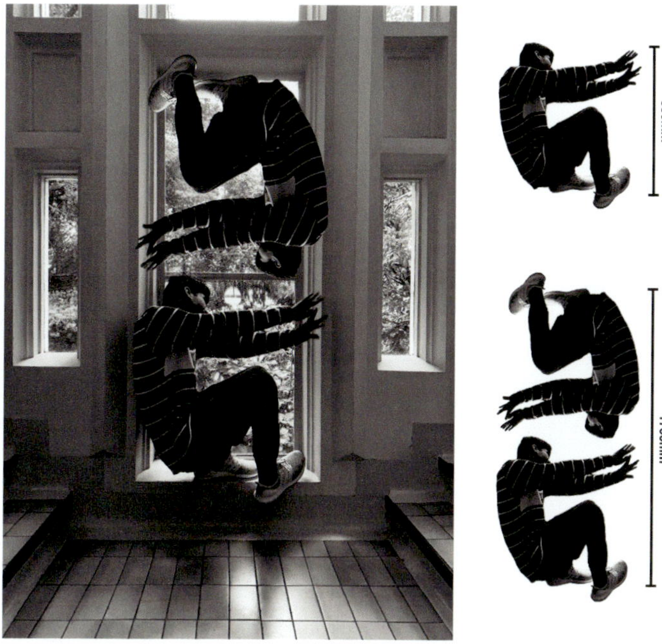

Figure 5.8 Architectural Media 2: Task 1, 2020, Samantha Bell.

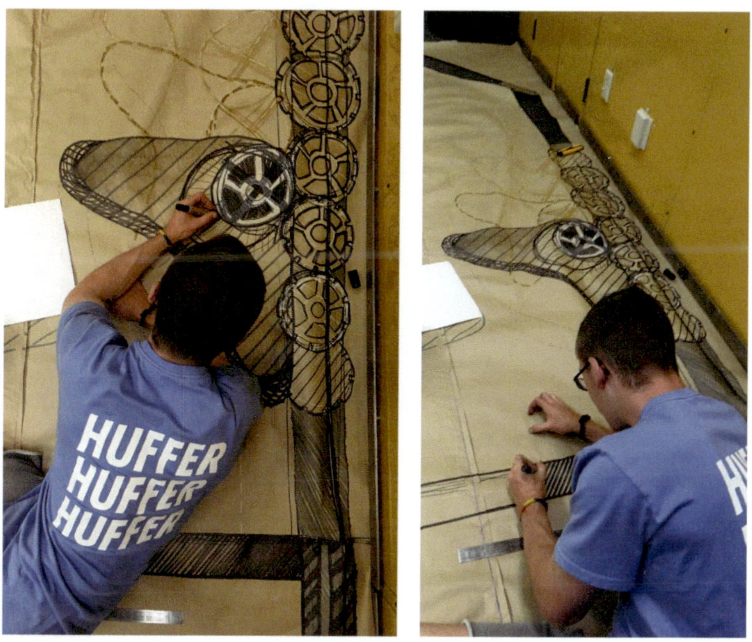

Figure 5.9 Architectural Media 2: Task 10, 2019, Sam Jones. Photograph by Pasifique Shabantu.

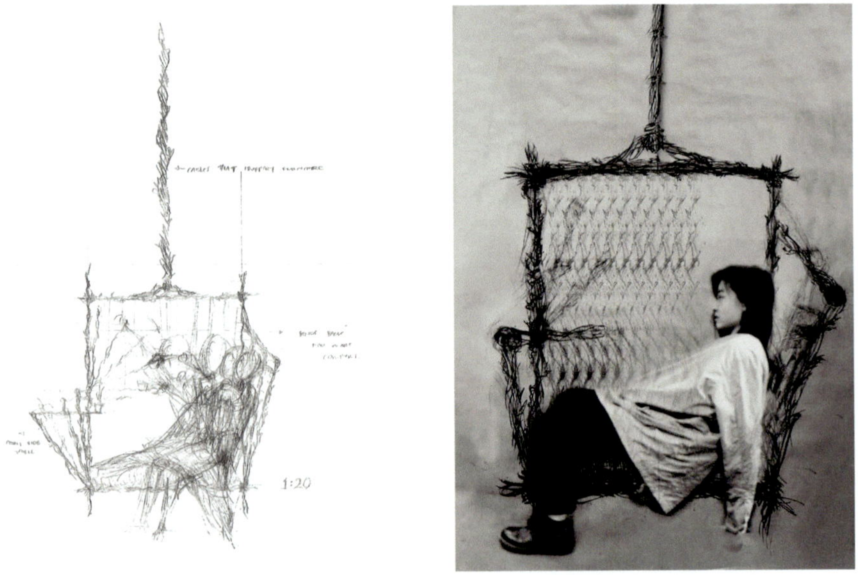

Figure 5.10 Architectural Media 2: Tasks 9 and 10, 2018, Ming Ming Pua.

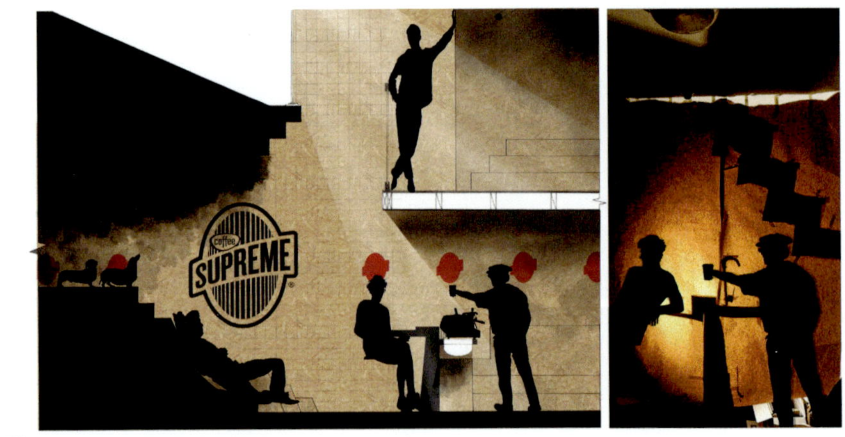

Figure 5.11 Architectural Media 2: Tasks 9 and 10, 2018, Sam Julian.

Drawing production becomes an enjoyable, dynamic interaction with peers. The students' own body is at the centre of the learning which resembles an enculturation or habituation to practice, an induction into the *questions* and *practices* of design.

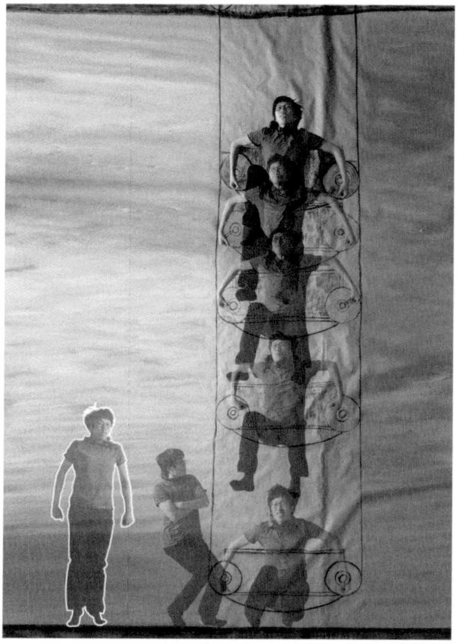

Figure 5.12 Architectural Media 2: Task 10, 2018, Daniel Ho.

Knowing time

In the programmes within which we teach, postgraduate degrees are research degrees. After acquiring disciplinary knowledge and situating themselves within the field in their undergraduate degrees, students progress to inquiring and understanding through design-led practices. The paradigm of creative-practice research – i.e. not only research *about* or *for* creative practice but also *through creative practice methods* – has flourished in the last decade, particularly in the UK, Scandinavia, Australia and New Zealand.[23] In architecture and related fields, master's- and doctoral-level research require students to identify topics or questions that will be of wider relevance than the student's own professional and personal development. Exhibition is a key method by which the work is developed, presented and evaluated (see Figure 5.13). In some universities, the exhibited work is recorded in a culminating written thesis; at others, the exhibition itself is taken to be the final word and is accompanied by a shorter exegesis providing context with a description of methods and processes. By focusing on inquiry rather than simulating professional practice, these degrees foster speculation, experimentation and critique. In their adoption of exhibition and installation as strategies, they offer possibilities for performative, embodied

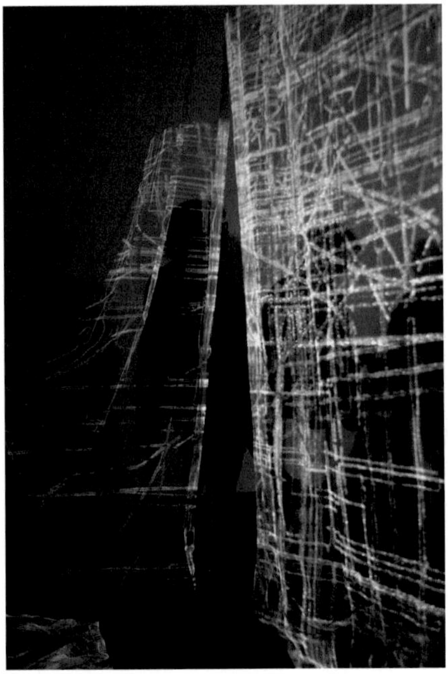

Figure 5.13 *Her Drawn Lines*, MArch(Prof) thesis project: installation view, 2019, Fritha Powell. Photograph by Dominic Wilson.

work. Two recent examples show how this surfaces critical questions about the nature and formation of creative knowledge.

In her Master of Architecture (Professional) research thesis – a year-long self-directed and supervised project presented within a dissertation – Sarah Barnes explores the body's presence in spatial drawing and the implications of this for perceiving time and duration. Her inquiry springs from the observation that design is *felt*: 'Like many other designers, the enjoyment I find in drawing plays a large part in my choice of pathway. The feeling of the pencil drifting across the paper, the lines and marks slowly accumulating.'[24] This is not a side note regarding the subjective pleasure of creative work but becomes a prompt to consider how the feeling of drawing might be a kind of architectural knowing. Through seven sequential sets of drawing, Barnes explores duration, endurance, repetition, constraint and orientation. Her drawings begin on the page but escape it, extending into walking exercises mapping landscapes. Throughout her dissertation, Barnes contextualizes her inquiry amongst other twentieth-century performative artworks such as Yves Klein's (1928–62) Anthropometry paintings, Matthew Barney's *Drawing Restraint* and Richard Long's walking sculptures.

First, she traced her own body on large sheets of paper, recording movement and changes of posture. In *Transitioning from Standing to Crouching (71*

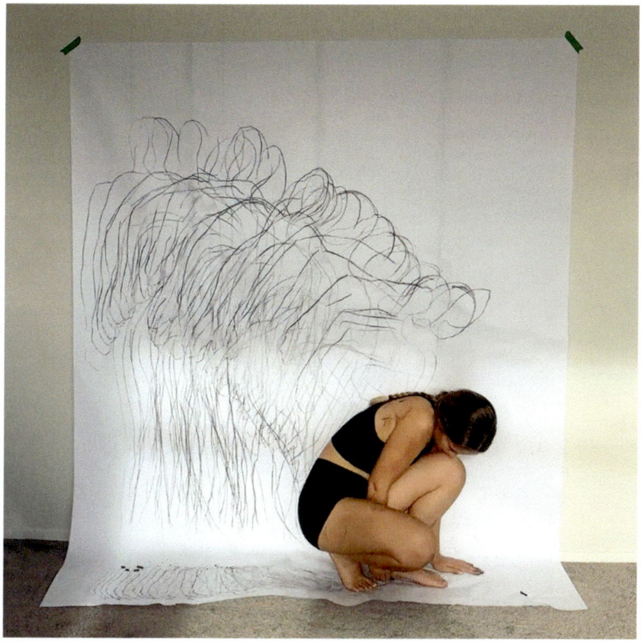

Figure 5.14 *Transitioning from Standing to Crouching (71 minutes)*, 2022, Sarah Barnes. Charcoal on paper, 5 55/64 × 7 37/64 in. (1800 × 2250 mm).

minutes), she began upright with her back against the wall and finished crouching alongside it (see Figure 5.14). As both the tracer and the form being traced, this required contorting herself and holding her body in demanding, physical positions. Narrating her experience in the third person, she notes '[t]hough her body is moving through space and time, the marks made allude to stillness'.[25] In Pliny's telling of the origin of drawing, Diboutades traces a static profile of her lover so that in his absence, his form remains. Barnes's self-tracings allow her to step outside herself and look back. Maurice Merleau-Ponty (1908–61) famously observed that when we put our hands together, we do not have two experiences of touching but a single experience: 'an ambiguous organization' of touching and being touched.[26] In self-tracing, the body is experienced from both inside and outside, simultaneously agent and object.

Barnes then moved on to experiences of physical endurance. Some acts of drawing were carried out until overcome by exhaustion or were counted in the language of physical exercise – in laps or circuits – or incorporated calisthenic positions like V-sits or wall-sits[27] (see Figure 5.15). In *Seeing How Long the Body Can Sustain Drawing For*, Barnes persisted drawing with a stone on the floor for five hours and twenty-four minutes (see Figure 5.16). Time, number of strokes and surface area all featured as challenges or limits. These drawings focused her

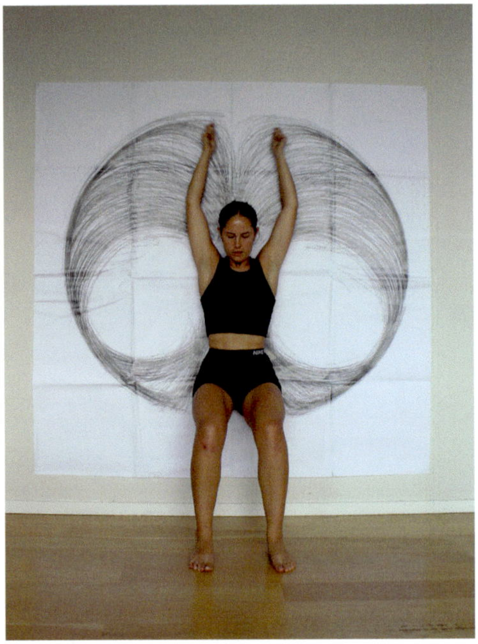

Figure 5.15 *Drawing While Holding a Wall-Sit, Until Exhausted*, Series 3, 2022, Sarah Barnes. Graphite on paper, 71 × 71 in. (1800 × 1800 mm).

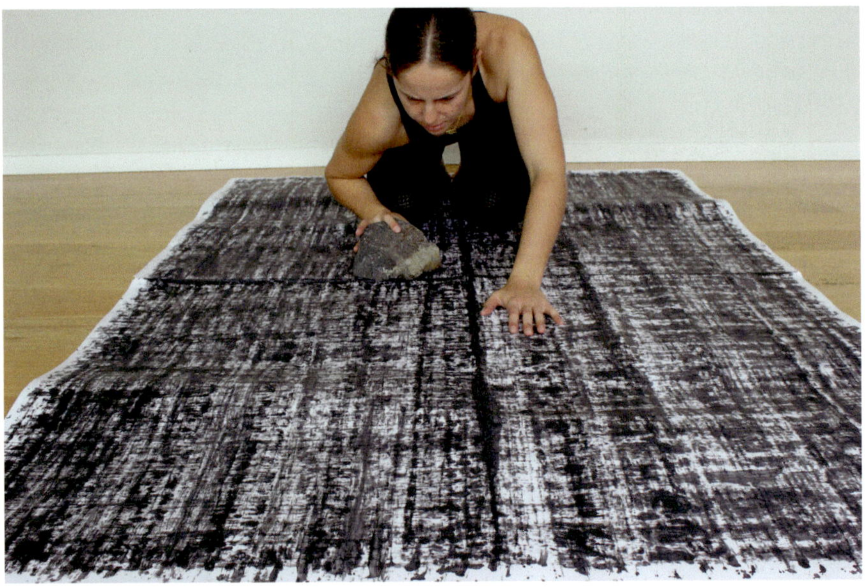

Figure 5.16 *Seeing How Long the Body Can Sustain Drawing For*, 2022, Sarah Barnes. Paint on card, 31 x 51 in. (900 × 1300 mm).

attention on rhythm and pace: strategies for managing her body. They dissolved the impulse for invention or the need to plan or think ahead, leaving only persistence. Through repetition, the particularities of each movement recede into an averaged gesture, a record of limits. A series of walking-drawings followed: journeys recorded by GPS trackers, made on moving trains[28] and paced out on beaches at low tide (see Figure 5.17). In *Drawing on the Landscape (57 minutes)*, she walked up and down to demarcate a square area of sand (see Figures 5.18 and 5.19). These works aimed to disclose or reveal rather than add; J. Ferreira Crespo writes, 'the first gesture of architecture is previous to any construction, rather architecture starts from an awareness of spaces and territories, done firstly by a recognition of what surrounds us and where we are. In this sense, walking might have been the first action to grant us that awareness.'[29] Once again, endurance is crucial. As Barnes's drawings extend to the size of the landscape and city, their scale is set by repetition and stamina: a sequence of bus stops, turnings or the placing of one foot in front of another.

Barnes's drawings address architectural themes of occupation, time, space and scale. She ties these to the physicality of drawing – what it requires of, does to and feels like for her body. While the drawings are definitively personal, they invite us to question assumed links between authorship and intention. Authorship in this case references fluctuating emotions, fatigue and limitation.

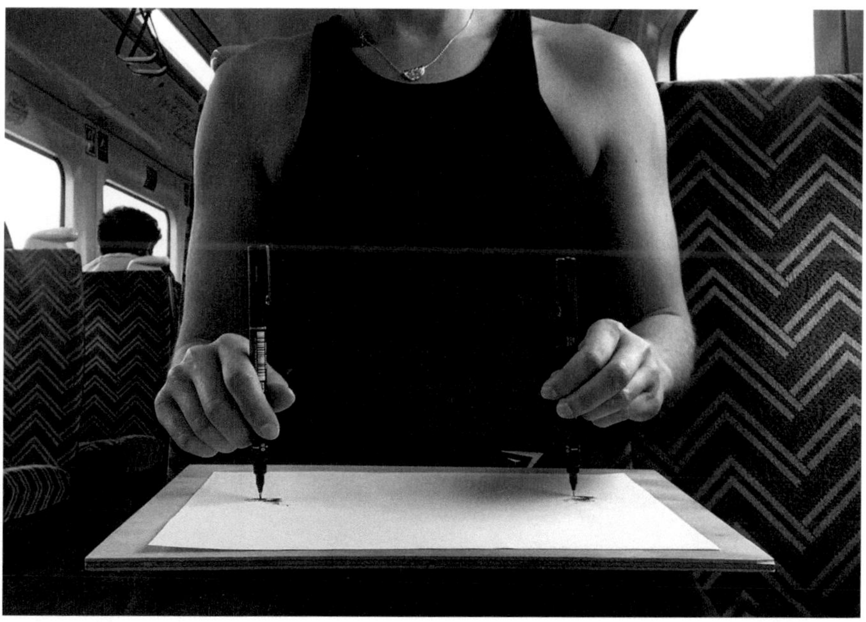

Figure 5.17 *Drawing While on the Train: Time Between Stops*, 2022, Sarah Barnes. Pen on paper, 8 5/16 x 11 3/4 in. (210 × 297 mm).

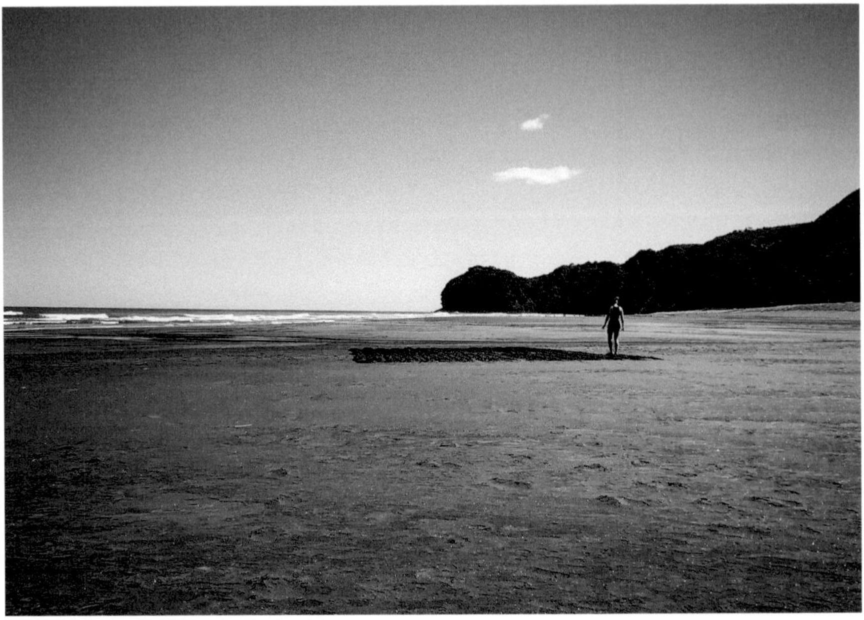

Figure 5.18 *Drawing on the Landscape (57 minutes)*, 2022, Sarah Barnes. Feet on sand.

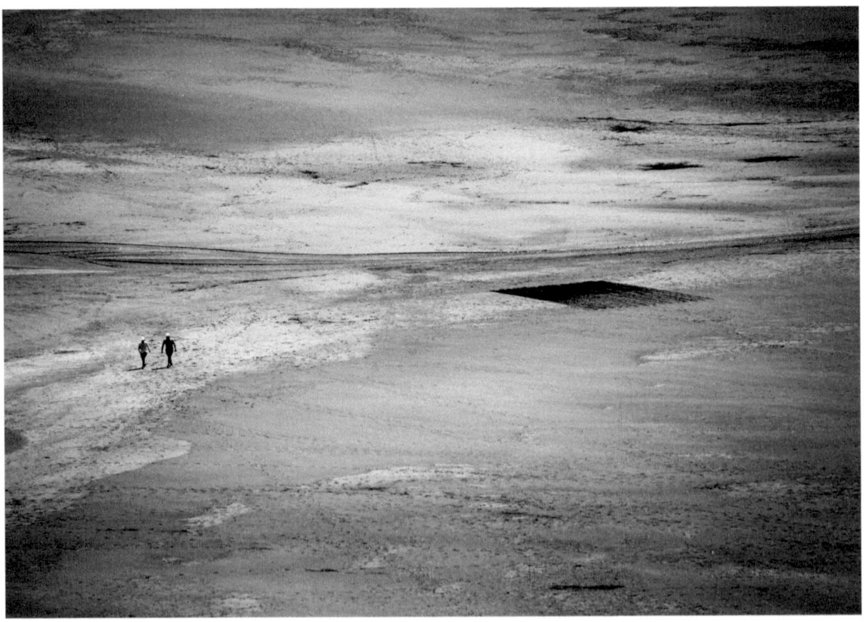

Figure 5.19 *Drawing on the Landscape (57 minutes)*, 2022, Sarah Barnes. Feet on sand.

Loss of control paradoxically reveals personal conditions. As Barnes puts it, '[these] lines extend beyond my control, affirming my existence and reassert a sense of self into the work . . . enabling the body to be entangled in the practice of drawing'.[30] Viewers vicariously re-encounter time through the drawings, either in their graphic residue or in watching them performed. The drawings gain meaning as they are imaginatively re-inhabited. One of Barnes's examiners wrote:

> The time felt by the body – through pain or exhaustion – does not have the dependability of conventional time where one minute is followed by an equal other. Bodily time might seem to lull or endure or, under different circumstances, disappear with lightning speed. This process of drawing brings the labour involved in its creation to the fore – a kind of labour that mimics other hidden repetitive work of, for example, maintenance or motherhood – that is experienced differently to linear, conventional time.[31]

Barnes's research inquiry does not produce generalizable abstract knowledge. Rather, it concerns the way learning is bodily situated, like a kind of muscle memory or tacit contextual understanding. Reflecting on her drawings, she writes of her body as 'an active database' that will condition any future design activity.[32] Architectural learning is as much a matter of bodily conditioning and tuning as an intellectual pursuit.

Barnes's inquiry stripped the act of drawing back to a minimum. Most of her drawings used little more than a graphite stick. Another recent thesis, 'The Learning Anatomy: A Performative School of Architectural Bodies' by Nicola Chang (2022), explored more elaborate drawing instruments.[33] While drawing instruments are typically thought of as a kind of tool, Chang sees them as prosthetics or costumes (see Figure 5.20). A costume comes with a persona, an imagined character, and implies the existence of a theatrical ensemble: an audience and perhaps a stage set. Her wondering what might happen if a design studio is imagined as a theatre or dance floor is a catalyst for the work, situated within performative artworks such as Nikolaus Gansterer's research project 'Choreo-graphic Figures', Rebecca Horn's (1944–2024) bodily prostheses and Lucy McRae's technology enhancements.[34]

Chang begins from the underlying assumption that architectural designing employs a framework of ideation, development and refinement towards a resolved spatial proposition. She wonders, however, about the 'hidden complexities of experimentation, problem solving, and critique' concealed in this model which aid the notion of design processes as mysteriously uncertain.[35] Chang writes: 'The true mystery in architecture school is uncovering how an architect performs their thinking.'[36] She notes that this uncertainty can seem intimidating for a design student yet important to foster as it opens into unrecognized, hidden opportunities in design. For Chang, the body is a source of productive uncertainty

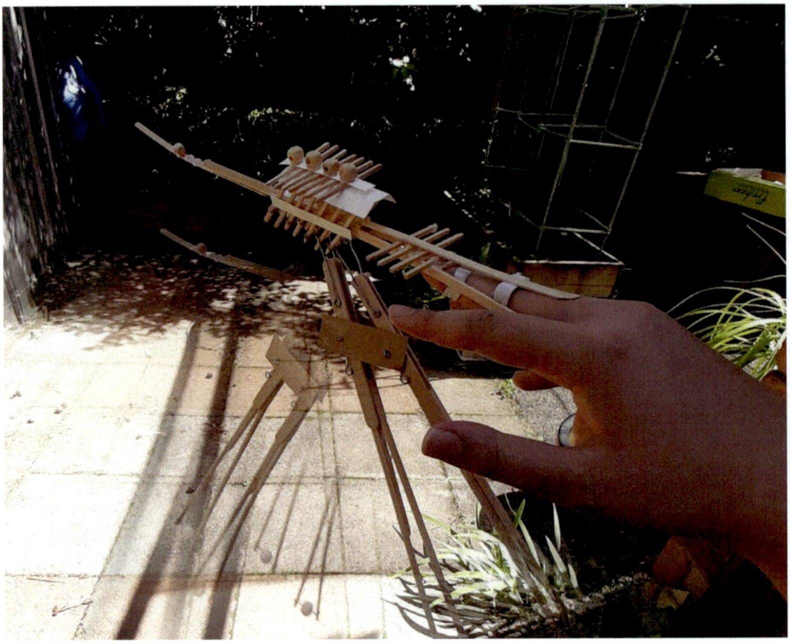

Figure 5.20 The Learning Anatomy: *Swinging Strokes*, 2022, Nicola Chang. Image by Nicola Chang.

that her experimental methodologies aim to exploit; if thinking is a bodily activity, how might dressing or equipping a body differently allow new thoughts?

Chang designed a group of characters, with specific costumes for each so that the wearer would embody the full scope of the character. Each of these characters is aware of their surroundings and role within the design process. The *Spectre*, for example, wears a tutu-like tent skirt, a headdress with pockets and a corset (see Figure 5.21). Their appearance recalls the flurries of accumulated paper sketches that might arise from brainstorming. *The Spectre* accumulates without judging, prioritizing fertility and multiplication rather than selectivity. With vision partly obscured, and constrained by the wide-reaching skirt, performers had a sense that they were an expansive presence: 'stately and grand', defining space as much as occupying it.[37] The pair, *Noughts and Crosses*, wore matching ensembles consisting of sleeveless shirts with attached cupboards (see Figure 5.22). These contained yarn dispensers in six colours which functioned something like talking sticks as they were passed from one participant to another (see Figure 5.23). They connected individual contributions into threads of conversation and fostered simultaneous lines of inquiry. Debate, as mediated by *Noughts and Crosses*, may have begun as a one-to-one argument but soon opened out into a productive tangle as positions and options multiplied. *Multi-Fool Eyes* was dressed in a vest with a kind of shingled surface that, when the torso spun, raised up at right angles to the body like winged windmills (see Figure 5.24). This was seen as

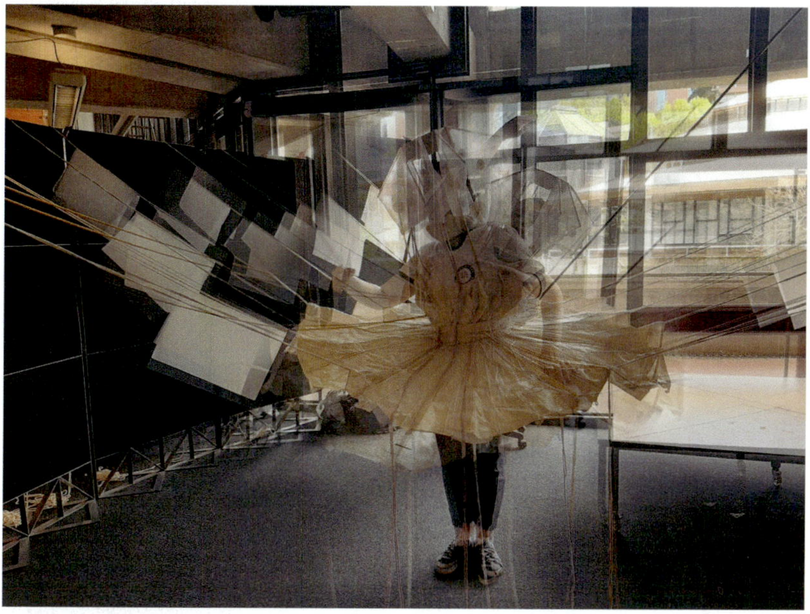

Figure 5.21 The Learning Anatomy: *Spectre*, conceptual collage, 2022, Nicola Chang. Image by Nicola Chang.

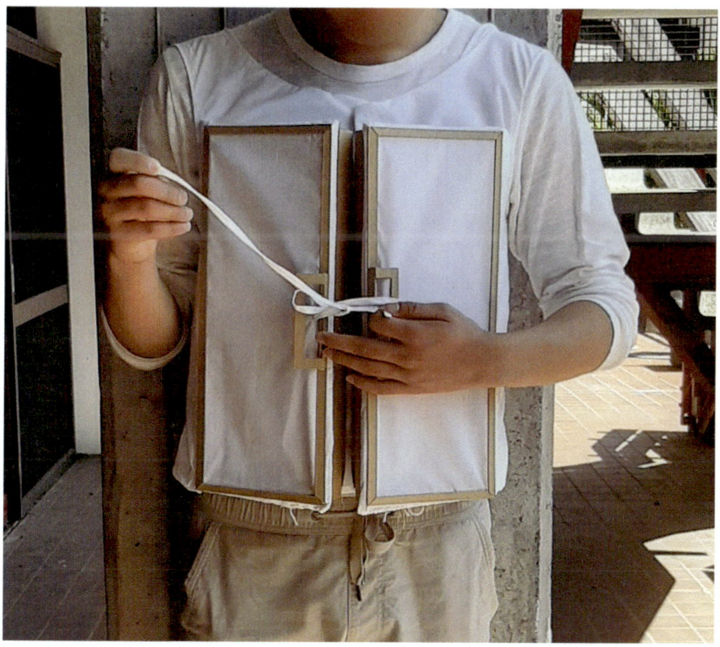

Figure 5.22 The Learning Anatomy: *Noughts*, twin cupboard shirt opening doors, 2022, Nicola Chang. Image by Nicola Chang.

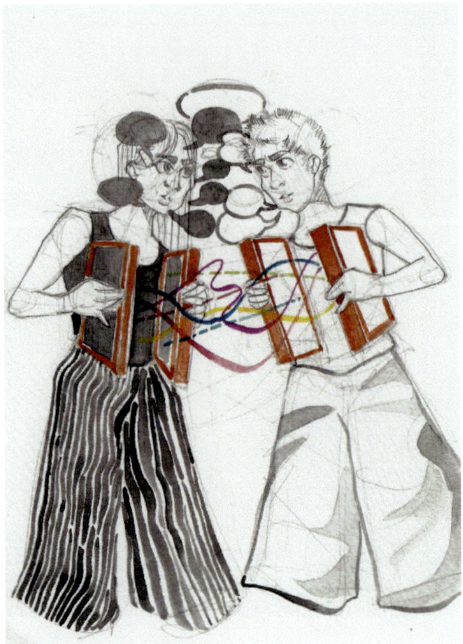

Figure 5.23 The Learning Anatomy: *Noughts and Crosses*, four character sketch, 2022, Nicola Chang. Image by Nicola Chang.

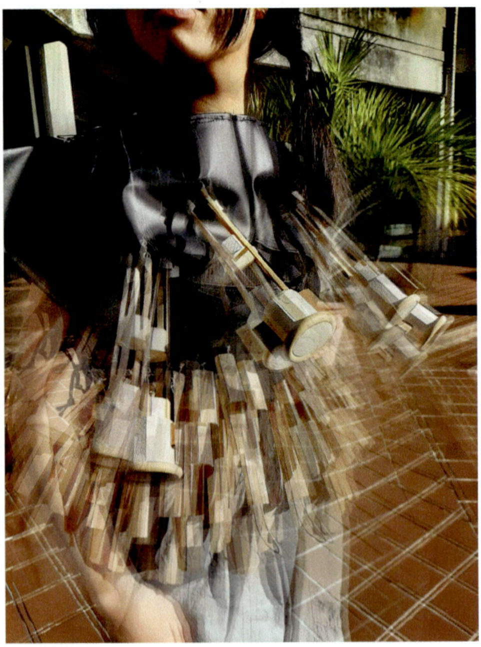

Figure 5.24 The Learning Anatomy: *Multi-Fool Eyes*, performer photo, 2022, Nicola Chang. Image by Nicola Chang.

the disruptive force for a seemingly resolved design – the creative critique that interrupts design thinking.[38] Counteracting the idea that a project necessarily comes to an end, it was intended to prevent closure. By operating within the space of the presentation, *Multi-Fool Eyes* 'ventilates this density, flinging drawings in an exaggerated manner' in order that the designer may discover new ideas within this mayhem.[39] Confusing the tidy order of presentation, the proper relationships between drawings and individual authorship, the character reveals new accidental possibilities latent in collective activity. In place of the intellectual phases of ideation, development and refinement, Chang proposed (in personified form) spectral accumulation, elaboration and collapse. These characters, like medieval fools, mask critique under performative humour and point to the importance of irrational or disjunctive leaps in designing.

Chang's characters facilitated a series of well-received workshops with other students (see Figures 5.25 and 5.26). They led design exercises that collectively structured and stimulated discussions of creative process across and between individual projects (see Figure 5.27). Chang orchestrated a carnivalesque studio of contradiction, uncertainty and ambiguity. Thinking, she proposes, is not a singular, intellectual activity, but might involve collaborative play, allegorical nonsense and divergent development. These were clearly not intended to be efficient ways to design. Rather they were extravagant, seeking to harness

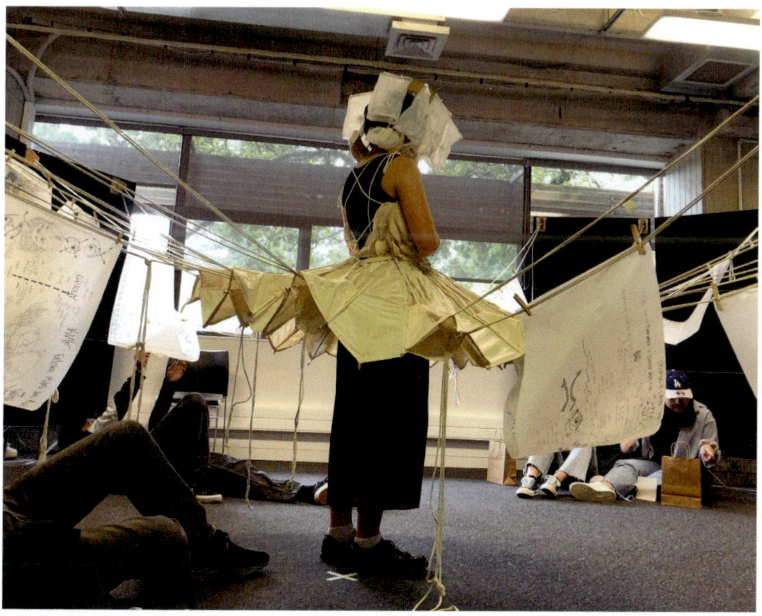

Figure 5.25 The Learning Anatomy: Activity Two, participants wandering and lying down under the tutu tent of sketches, 2022, Nicola Chang. Photograph by John Golena, edited by Nicola Chang.

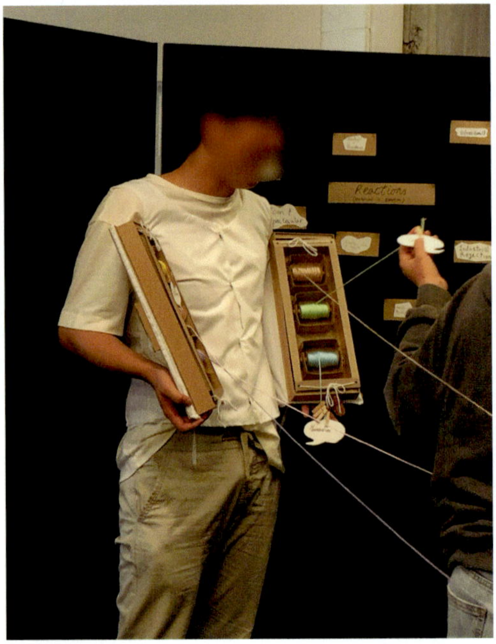

Figure 5.26 The Learning Anatomy: Activity Three, participant picks a coloured yarn, 2022, Nicola Chang. Photograph by John Golena, edited by Nicola Chang.

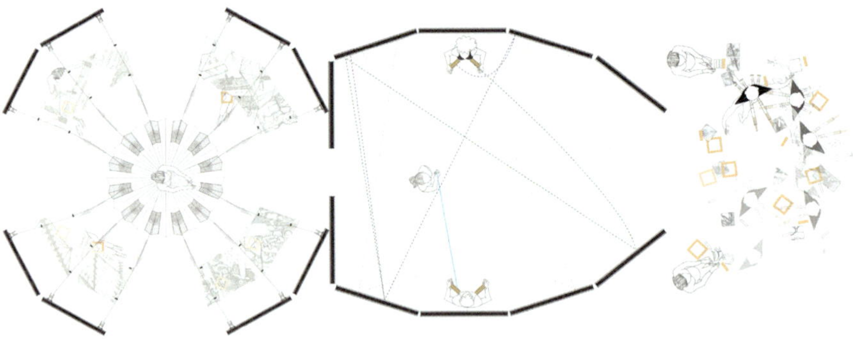

Figure 5.27 The Learning Anatomy: A Performative School of Architectural Bodies, 2022, Nicola Chang. Image by Nicola Chang.

creative value from reconfiguring the corporeal space and procedures of the studio. Like Barnes, Chang is fascinated with the way designing *feels* as a bodily activity: *performing* design thought. For both, a full-scale drawing practice opens an inquiry into the nature, activities and times of the studio.

Discontinuities

At various times, design will include production, hesitation, endurance, interruption, deliberation, conversation and play. The design process has been modelled in various ways but is commonly described as having stages, phases or steps organized into lines or loops: a collection of connected activities. In Chang's performative model, these become characters. What is carried over from one phase to the next? What translation and transformation occur at the thresholds between them? 'Object-Sketch' (2024), a Master of Design thesis by digital fabrication researcher Matthew Davis, explores the materiality of the line as it moves between gesture, surface, software and three-dimensional objects.[40] In his interactive installation, lines, defined by ropes hanging on a wall, were encoded into generative software, then converted into toolpaths for a digitally controlled ceramic printer. This 'toolchain' was not intended as a practical or efficient means to an end but as a demonstration of a non-linear design method.[41]

Sketching, as we have already seen, is typically associated with conciseness and provisionality. Sketches are typically overtaken by more definitive modes of drawing as a design process moves towards stages of resolution and realization. Davis, however, asks how the design process might remain sketchy throughout. The sketch as he defines it is 'a mediated meandering'.[42] It need not be surface-bound or hand-made: in his terms, a digital model, an algorithmic description or a physical prototype can all be sketches insofar as they have a certain looseness or approximation: a preliminary quality. His sketches are multiple; no single line is authoritative. What differentiates these various modes of sketching is *materiality*. Materiality is qualitative and manifests particularly in irregularities, inconsistencies, characteristic errors or glitches. Sketching, according to Davis, thrives on this resistance. Provisional and flawed, it describes not one thing but a range of possibilities. When the sketch is translated, these open up: 'Eddies and moments of pause in the transcoding between processes and tools might reveal new potentials and paths for actualisation.'[43] Davis establishes a practice of tinkering, of chaining processes together in ad hoc, sometimes unorthodox or deliberately awkward ways, to reveal the displacements, transformations and frictions of materiality. Davis observed that the rhetoric associated with computer use is often about the absence of friction and the promise of acceleration. By contrast, Davis's toolchain for digital fabrication was something like a Rube Goldberg, or chain-reaction, contraption. At each stage of the process, the sketch was re-abstracted and re-actualized in new ways. These could easily have been reconfigured (and were in previous iterations): the data could have driven a different fabrication machine; the algorithm could have transformed the data differently; the physical interface could have been assembled in other ways.

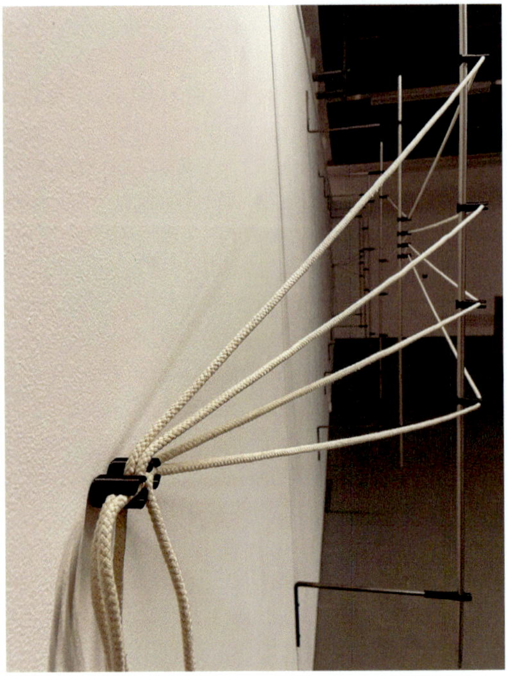

Figure 5.28 Object-Sketch: Interface 1, 2024, Matthew Davis. Photograph by Matthew Davis.

As exhibited, the work included a large drawing machine: a physical interface consisting of ropes hung from metal bars by sliding clips (see Figure 5.28). These could be repositioned by visitors to form a series of profiles. The system consciously referred to the historical practice of using long, thin timber battens to shape the hulls of boats. (These were known as splines, the origin of a now-common digital term.) It also recalls Albrecht Dürer's (1471–1528) perspective constructions and Leon Battista Alberti's (1404–72) scanning and surveying techniques (see Chapter 2). The system was constrained: there were a limited number of ropes and they could only attach to the bars in certain ways. Slack in the ropes prevented them from forming straight lines. Changing the lines, moving the sliders, clipping and unclipping ropes was a haptic experience. Given the size of the apparatus, participants tended to stay close to make adjustments and then step back to view the overall arrangement, mediating through their bodies the relationship between detail and whole (see Figure 5.29).

Sensors in the sliding clips captured their position as data, which was carried by a loom of wires to a computer behind the scenes and visualized on a screen. The visualization did not directly represent the rope-lines but used their position

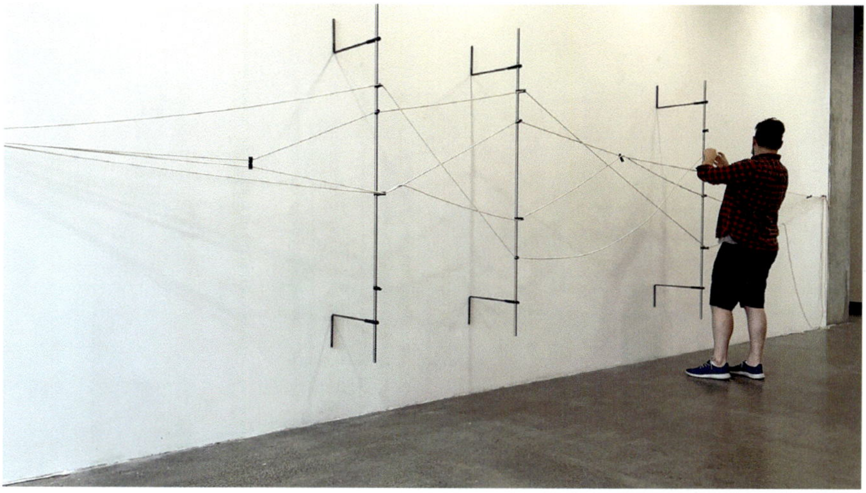

Figure 5.29 Object-Sketch: Interface 2, 2024, Matthew Davis. Photograph by Matthew Davis.

to generate a new animated digital drawing: a cluster of lines meandered up the screen, coordinated as a simulated flock.[44] The slider positions supplied a gravitational or magnetic force acting on the flock. The lines were redrawn every few seconds, varying randomly based on the state of the rope interface. Davis selected sets of lines which were then converted into tool paths for a custom-made ceramic 3D printer. Slowly, over the course of hours, the printer extruded a fine bead of clay in layers to produce a unique vessel. The printer, too, was a drawing machine, slowly making its spiralling line in space (see Figure 5.30). The ceramic vessels, once fired, were exhibited in the gallery opposite the rope-line interface.

'Object-Sketch' treated design as a process of linking together processes drawn from an open-ended repertoire. It was, as Davis describes, 'a tool for transcoding movement . . . bringing our gestures into conversation with a coded interface'.[45] Sketches were not only a means to initiate a design but also the model for how a design – at all stages of the process – both narrows and opens, clarifies and renders ambiguous. The knowledge produced by Davis's thesis is not how to make a particular kind of ceramic vessel but how the design process can be constructed as a series of actualizations and abstractions. Barnes strips away supporting equipment to foreground the bodily labour of drawing, while Chang stages drawing as a social and corporeal event to reveal narrative labour. Davis's drawing machines frame the design process as fundamentally sketchy.

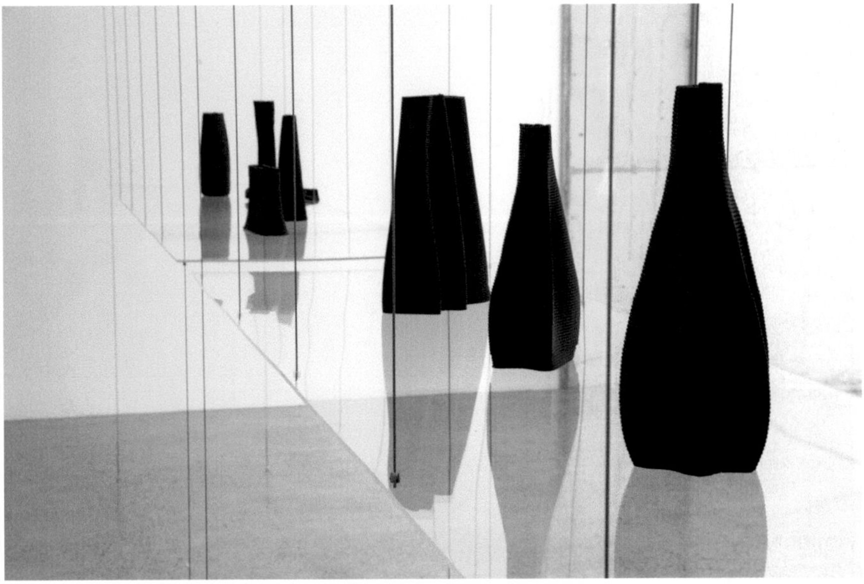

Figure 5.30 Object-Sketch: Interface 1, 2024, Matthew Davis. Photograph by Stephen Cleland.

Making studio work

Each of these examples treats the studio as a site of inquiry. This is important for learners orienting themselves within design to understand and remains an important aspect of a reflective practice. The studio is a place of ongoing practice rather than discrete projects. While structured approaches to design methods have value, it is also vital to recognize the role of improvisation, circumstance and feeling. These '[t]actical improvisations accumulate over time to produce new models for operation', becoming a repertoire for future action.[46] These contingent and particular aspects arise from situated bodily activity – design that happens in particular places, has its own temporality and centres on the body.

There are, of course, many examples of inventive studio teaching that foster an understanding of design as an improvisatory repertoire. One of these is outlined by Elizabeth Hatz in relation to the process of a thesis student, Isak Boardman, whose work she supervised at KTH Royal Institute of Technology, Stockholm. During design development, the inevitable shift to digital drawing seemed to be to the detriment of Boardman's creative process; her advice was to maintain a physical drawing practice.[47] This Boardman did, by laying out A4 sheets of paper – as substitute for the proposed building's thick walls – on the ground of the site under consideration. This allowed Boardman to work directly

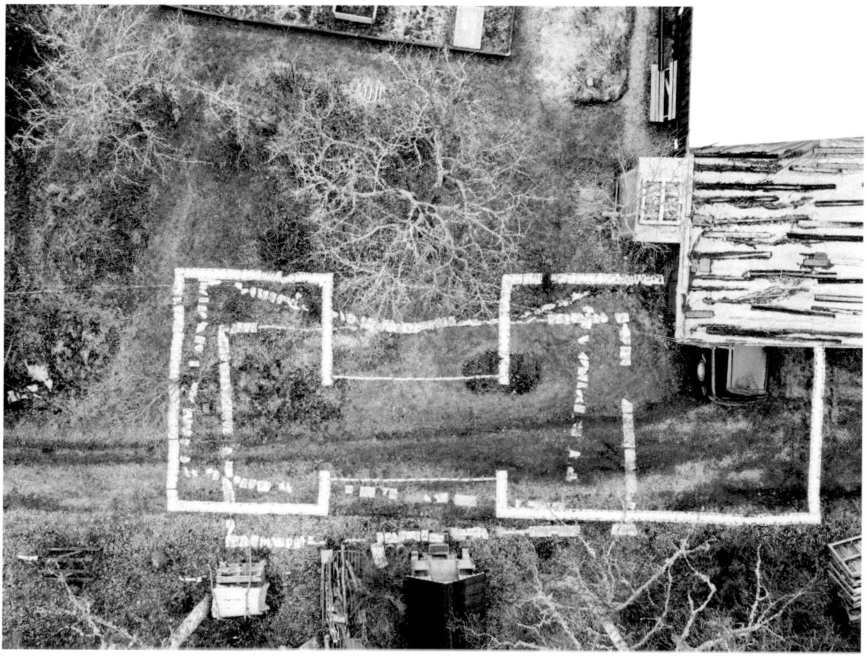

Figure 5.31 *House for Father and Son*: Sista A4-sketch, Isak Boardman. © Isak Boardman Architecture.

with the site, adjusting his intentions by taking in particular aspects of it. A drone was then employed to photograph the resulting in situ plan (see Figure 5.31). Boardman writes of his intention to find architectural drawing tools that could work as naturally as an extension of the body.[48] These vulnerable sheets of paper on the ground, easily affected by wind and rain, were likened to the fragility of a pencil sketch.[49] Maintaining this embodied drawing practice enabled Boardman to explore space as something immediately present rather than only belonging to an imagined future realized through visual or schematic representation.

In another example by Eva Flores and Ricardo Prats, graduating students at the International University of Catalonia were asked to design a container to store everything they produced over the final year of their degree. This included all process work, models and drawings. The container was to be portable and store the work securely (see Figure 5.32). For the final presentation, students were to open the container and unfold their creative journey by taking out and narrating the objects contained in it:

> The order in which the material was laid out helped the students to remember what they were going to explain, their final presentation became a performance more than a simple explanation of drawings or panels: as they opened up the

parts of the container, in line with the narration, the students had to move around, extend their arms and walk as they spoke. This converted the tension of the occasion into an expressive power that aided the physical explanation of the project.[50]

This treatment had previously been used by Flores and Prats in their own practice, which often involves full-scale drawings, installations and models.[51]). The cabinet, or case, invited students to consider what constituted the material productions of their practice and required reflection. It was an archive, organizing, attributing value to and storing work for future use, and a travelling case, implying relocations and journeys and a mobile stage. Students' attention is directed to the value of interim, provisional attempts, experiments and the partially resolved. Rather than evaluating how well they executed a pre-determined design method, they were encouraged to act autoethnographically, looking back on the things they made and did, discovering critical moments and reinventing them retrospectively.

It is notable that design (as an applied practice) is believed to have method or design process – a kind of logic that can be mapped, shared, optimized and progresses towards the moment of realization (whereas art is not typically expected to have such an underlying system). Studio teaching is commonly structured around this logic: a design brief leads to conceptual investigations with input from precedent studies. In a series of progressively more resolved, realistic

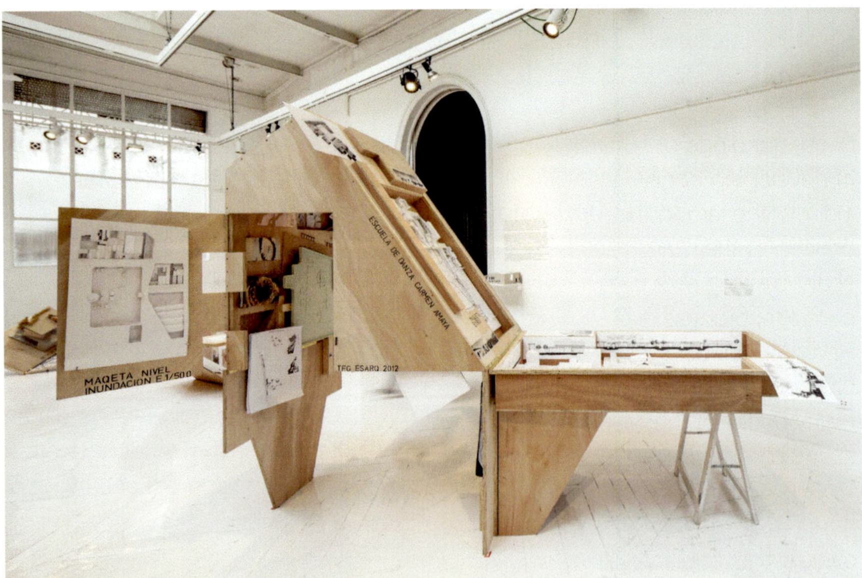

Figure 5.32 Portable Memories Exhibition, Flores & Prats, 2013. Container by Joan Sanz. Photograph by Adrià Goula, courtesy of Flores & Prats.

and specific iterations, the proposition makes its way towards completion (or at least the final project review). Varying this sequence and offering students the opportunity to explore other ways to give narrative shape to a creative process is important. These postgraduate design research theses discussed each inquire into the structure and experience of creative practice. Drawing, which is often used to reinforce a linear idea of design process, instead offers new temporal structures.

In Chapters 1 and 3 we discussed experimental student work from the Cranbrook Academy of Art. This small but influential post-professional architecture programme, established in 1932, does not have a fixed curriculum nor aim towards professional accreditation. Instead, students undertake studio-based research that emphasizes physical making. Gretchen Wilkins writes, 'Making also embeds a quality of perpetual self-propulsion into the work, and by extension into the practices that emerge from the process. The work itself generates questions that provoke new ideas, deeper reflection or further collaboration, in itself also an engine to sustain experimental approaches to practice outside the unscripted and permissive environment of the Cranbrook Academy.'[52] Unscripted does not mean aimless or without rigour; on the contrary, students themselves are required to establish aims, organize their activities thoroughly and strategically and frame the work's value. Emphasis on physical making and the specificities of physical situations entails critical reflection on the activities of practice. As Temple writes: 'Learning design by immersion in concrete experience through direct contact with materials, purposefully for discovery of a material's workable properties, enables development of critical capacities about design decision making.'[53]

Corporeal pedagogies encourage students to recognize the agency of their bodies in learning. From early in design education, they enable better understandings of space, situation and human relationships. They aid students in developing bodily intuitions of scale that go beyond metric comparison. They include the embodied and situated knowledge and experience that students bring with them and facilitate their interpretation of their existing contexts. Full-scale drawing as a corporeal pedagogy opens a space of inquiry in which aims, contexts, process, protocols for interaction and values are yet to be determined. Rather than being 'a horizontal or vertical slice in one moment in time' as orthographic drawing is, full-scale drawings mediate time in rich ways, exploiting the capacity of the drawing to be both a portal to another place and time and a situated enactment in the present.[54]

The studio is a place of situated inquiry. Although it might at times take the form of a simulation of practice, it cannot be directly conflated with the professional office. It is a place for inquiring into indeterminate, complex or ambiguous conditions, not merely a place to solve or resolve problems into projects. According to Peter B. Hyland and Tyson E. Lewis, studio 'is not a retreat from the world, but rather a suspension of the world so it can be studied';[55] studio (at least in principle) foregrounds contemplation and experimentation. One of the basic premises of studio pedagogies is that teachers do not simply impart information and skills; rather, students and educators face this shared uncertainty together.[56] Ultimately, decisions must be taken and acted upon without any prior guarantee that they will be correct or constructive. Ongoing reflection on this path of choices enables learners not only to identify how they might respond to similar specific situations in future but also to develop strategies for inhabiting uncertainty.

CONCLUSION

THINKING THROUGH DRAWING

Thickening drawing

Architecture has a consequential and entangled relationship to drawing and one which is continually in flux. According to the Renaissance separation between designing and building, designs are made by drawing and these drawings mediate between the studio and construction site. Recently, these mediating drawings have expanded to overlap with models, simulations, visualizations, maps, diagrams, databases and code – to the point that other terms like *media* or *image* might seem more apt. In this book, we have particularly explored the expansion of architectural drawing into installation and performance modes. Despite these changes, however, we retain the term *drawing* to emphasize continuities and transformations of practice rather than breaks and disruptions.

A key assumption, arising from this Renaissance separation, is that architectural drawing is a relatively transparent and referential means of communication that exclusively points towards, serves and culminates in acts of building. By now, it is well-established that this is not how architectural drawing works. Architectural drawing refers outside itself, but it is not *purely* referential. Developments in drawing theory, at the end of the twentieth century, championed this shifted relationship to reference.[1] Drawing's non-referential aspects – its *opacity* or *thickness* – are vitally important to its imaginative and sense-making operations. Just as, according to Paul Emmons, the rational and embodied modes of drawing are not exclusive but describe two tendencies always at work in architectural drawing, we see transparency and opacity, or thinness and thickness, as creating productive tensions.[2]

The rich array of recent publishing demonstrates the vigour of contemporary architectural drawing theory. These include themes such as architecture's interweaving with contemporary image culture (*Reality Modeled After Images: Architecture and Aesthetics After the Digital Image*) and visual art (*The Colors of Rhetoric: Places of Invention in the Spatial Realm*);[3] the lives of drawings and models after construction (*The Routledge Companion to Architectural Drawings and Models: Translating, Collecting, Archiving and Displaying*) and through dissemination (*Drawing Attention: Architecture in the Age of Social Media*);[4] overviews of contemporary drawing practices and production by architects, critics and curators (*Drawing Architecture: Conversations on Contemporary Practice*) and expanded practices (*The Artful Plan: Architectural Drawing Reconfigured*).[5] Journals, such as *OASE*, *Drawing: Research, Theory, Practice*, *Interstices: Journal of Architecture and Related Arts* and *DMJ–Architecture and Representation*, regularly include significant articles on these topics; *Drawing On: Journal of Architectural Research by Design* and Routledge's *Design Research in Architecture* book series demonstrate the importance of creative practice design research which are core to drawing and installation. In this light, even the recent relocation of the Drawing Matter archive, from Somerset to the centre of London, may be read as significant.[6] These contemporary currents make clear that the legacy of the flurry of publishing on architectural drawing in the 1990s is strong and ongoing. There are many significant contemporary practitioners of architectural drawing who we have not discussed, for example: Brian Cantley, Perry Kulper, Nat Chard, Penelope Haralambidou, Mark Smout and Laura Allen. Readers are likely already aware of other important contemporary and historical figures or sources we have omitted. In general, our strategy has been to prioritize less well-known works, and, more specifically, operate within a subset of architectural drawings which we refer to as full-scale drawings.

As we use it, the term *full-scale* does not predominantly refer to a drawing's size nor its metric proportionality; a drawing can be full-scale in our sense even if very small or when it includes an explicit proportional scale. Full-scale, for us, means that the drawing's frame of reference is provided by the presence of human bodies; that is, it addresses the space and time of the viewer (or the drawer, user, participant), rather than suppressing it. Scale, in the sense of proportional equivalence, entails correspondence and reference. The full-scale drawing refers to attendant bodies and to the here and now, yet also allows for reference to elsewhere and elsewhen.

Installation and performance art, body-centred genres that come to the fore in the 1960s and 1970s, turn on the idea of the artwork as a situation rather than an artefact. A situation has (and is) a context. It can be either a place or a happening defined in terms of relative orientation rather than an absolute frame. Situations entail the possibility of things occurring – of activity, action or use – and are therefore circumstantial and temporary. A situation is not locked to a specific scale; rather multiple, interconnected scales emerge through it. Situations may layer more than one place at a time, as we saw in Chapter 4. Full-scale drawing merges strategies from installation and performance with those from architecture, involving present-tense, situated and embodied aspects of drawing rather than the projective, anticipatory aspects typical of architectural drawing.

The *Drawing In* series, to which this book contributes, asks what it means to think through drawing. Thinking materially and bodily has the effect of thickening situations rather than rendering them transparent. While clarity, efficiency and immediacy may be desirable qualities, we point to the ways in which drawings thicken situations, slowing and deepening our experiences and understandings of them. The contemporary notion of the interface as a membrane through which action passes frictionlessly (think of the apparent immediacy of the touchscreen, as discussed in Chapter 2) contrasts with drawing's frictional interface that amplifies the transformative effects of translation. As a thick interface rather than a transparent window, a drawing can layer situations together; this is perhaps the distinguishing feature of architectural drawing.

A repeated theme in this book is the difference between two types of line. The schematic, conceptual line that leaps between two points (Ingold's transport line or Euclid's shortest distance between two points) is fundamental to architectural abstraction and exactitude.[7] It enables orthographic and perspective drawings and their integration into a system for modelling form in three dimensions. It also implies atemporality: the line is not imagined as travelling within the space between two points but is instantaneously present. The schematic line is reassuringly simple and easy to scale. By contrast, the movement-line (Ingold's wayfaring line or Klee's gestural line) must actually get from point A to point B, dealing with all the circumstances it encounters along the way.[8] The movement-line is concrete and particular, involving material conditions, context, living bodies and time. To scale a movement-line involves interpreting it and perhaps sampling or simulating it. Where the schematic line is defined in terms of two locations in Cartesian space, the movement-line creates space as it goes; David Rosand writes: 'To make a mark or trace a single line upon a surface immediately transforms that surface, energizes its neutrality; the graphic imposition turns the actual flatness of the ground into virtual space, translates its material reality into the fiction of imagination. Disrupting the emptiness, the mark activates the surface, disclosing dimensions latent in its suggestive blankness.'[9] Contemporary literature on drawing sometimes seems to talk past architectural drawing: in focusing on the living, temporal movement-line, the technical drawing is sidelined. Similarly, the artists to whom we refer have what can seem, from a designer's perspective, like the luxury of considering the drawing as an end in itself without the constraints of construction. For a designer, the drawing can never be purely this. It always, at least potentially, points elsewhere and elsewhen, containing seeds of future action. But even the most schematic, ends-oriented drawing involves body, movement and time; and the most circumstantial movement-trace can be abstract and projective. Architectural drawing, as we see it, negotiates between these possibilities.

Architectural drawing is already architecture, not only its future possibility.[10] Architectural thought is dependent upon drawing materials, techniques and instruments, but equally on physical situations and the bodies of drawer and viewer. Full-scale drawing, for us, is important precisely because it asks how these situated aspects thicken and enrich architecture. The dichotomies of abstraction and concreteness, projection and performativity, are not a problem to solve but instead enable productive tensions. While architecture anticipates future buildings, it can also have temporary, provisional, speculative, critical, retrospective past- or present-tense modes. These nuanced modes are particularly activated by full-scale drawing.

Expanding drawing

We began Chapter 1 by considering a particular historical moment. From the 1960s, new ideas about drawing interacted with emerging genres of art practice: installation and performance art. By the 1970s, drawing became understood as a spatial, situated, bodily activity in the present tense rather than primarily a means of picturing or representing. Concurrently, architects posited that the things they made – drawings and models – were already independent works of architecture rather than merely leading towards it; this was initially posed as a problem of how architecture could sit in a gallery context. The resulting exhibitions ultimately remained conservative and tended to lean on normative conventions of art to give architectural drawings and models credence and value as (saleable) art objects. This cross-pollination with the commercial art world lent weight to the idea that architectural drawings could be autonomous and disturbed what had (at least since the end of the nineteenth century) been a relatively settled graphic system. This encounter gave rise to a flourishing culture of speculation, active hybridization of drawings and models, interest in installation and exhibition as modes of architectural practice and critical perspectives on architectural representation. While it is true that computers transformed architectural drawing from the 1990s onward, an overemphasis on narratives of technological change can obscure this important disciplinary shift.

In Chapter 2 we looked back to the origins of computational drawing, discerning how these techniques embed full-scale drawing practices. Perspective drawing that simulated human vision, scanning techniques for mapping human bodies in three dimensions and sampling methods used in topographical surveys all involved three-dimensional procedures that we suggest can be seen as full-scale spatial drawing. Using cords, hooks, sightlines, rays of light, frames, screens, stakes, chains, graduated rings and other instruments, drawers constructed interactive three-dimensional drawings that mediated between a virtual Euclidean world of abstract geometry and the unruly material, corporeal world. Such drawings, we suggest, were forerunners of the pivotal contemporary concept of the digital interface. While computation disrupted traditional architectural drawing, it did not necessarily displace the human body. On the contrary, we suggest: it was founded on technical and imaginative strategies to bring the worlds of geometry and matter – mind and body – to coincide. These drawing strategies and techniques seem to offer a transparent interface. By seeing drawing this way – as an interface – we soften the distinctions between drawing-as-artefact, drawing-as-action and the instruments of drawing, that is, between drawing-as-means and drawing-as-end. Our implication is that the expansion of drawing, described in Chapter 1, activated latent potentials.

In Chapter 3, we returned to the late twentieth century to consider performative drawing. As art drawing in the 1970s flourished and expanded, the physical enactment of drawing was often given more importance than what was drawn. Even a single drawn line was understood to be a record of something happening in time, involving specific human bodies acting in particular places. Our opening examples dramatized the tension between ideal lines and bodily actualized ones. Rather than imagining lines as abstract projections from point to point, these works embark on a journey. They capture motion, expressing and recording a process in which the unconscious, corporeal, emotional and collective often overrode rational intention. We found sympathy between these intuitive, sometimes transgressive artistic lines and technologies of motion capture, beginning with the camera. In the twentieth century, architects became fascinated with the expressive potential of the gestural line: the line as embodied time. From the rising prestige of the concept sketch to the rigours of time-and-motion study to interest in flows and flowcharts, such lines came to imply the temporality of human bodies. Ultimately, this temporality was often stilled, treated as a means to generate static building forms. Drawings by Mulla, Paterson and Macken and by Manalopoulou and McLaughlin are offered as counter-examples that seek to avoid this freezing effect by using motion-captured drawings to explore how buildings are performed by their occupants. Closing the circuit between drawing and building, they steer away from assumptions about the completeness and closure of architectural projects.

From performance, Chapter 4 turned to installation and the possibility of an inhabitable drawing. Inhabitation is a kind of spatial investment and encounter that involves imagining, using, sharing and perceiving. It is a spatio-temporal encounter; and in this sense, drawings too can be inhabited. Once again, we look to the expansion of drawing in the late twentieth century, noting the emergence of installation as an avenue of architectural practice and inquiry. Installation, as a genre, activates the embodied presence of the viewer, occupant, participant or audience: it situates, creating a field of encounter instigated by viewers. It overlaps with exhibition but is not simply display or simulation. Our examples exploit the ambiguity of being both here and elsewhere, using strategies of scalar play, temporal and material resonance, and specifically engaging the liminal, barely perceptible and irresolute. Such creative practice research opens critical, speculative and reflective modes: not only critique *of*, speculation *about* or reflection *on* architecture but also critique, speculation and reflection *through* architectural means.

We believe embodied practices are particularly valuable for students of architecture because they foster these skills of critiquing, speculating and reflecting through drawing. In Chapter 5 we discuss a series of examples from our own teaching practices as demonstrative of how we consider the inclusion of embodied architectural drawing and what it enables for learners. Many architecture programmes no longer include core architectural media courses, on the assumption that these can be integrated into more general courses. The architectural learning studio, we argue, is not a mirror of the commercial design office, nor solely a place of content-delivery. Rather, it is for conversational inquiry and habituation to a range of practices, where learners develop a repertoire for acting in ambiguous, uncertain and changing situations. In this sense, studio is fundamentally embodied and situated (no matter what blend of online and offline situations it might involve). There are links, we argue, between the pedagogical deprioritization of the body, the atrophy of studio as a space of inquiry and the dislocation of the body in architectural representation. We suggest two stages where an embodied drawing practice can be valuable for students. Firstly, at the beginning of their studies, it can widen their understanding of field that constitutes architecture and counter unwarranted assumptions. Full-scale drawing can help students anchor learning in physical effects and relationships. As drawings, they retain provisionality and flexibility. As enacted material and corporeal performances in the present, they open imaginatively to other places and possibilities. Secondly, at postgraduate level, full-scale drawing can act as a place to critically reflect on their experience, knowledge and practice, and explore alternative structures of architectural practice.

Drawing open

Intentionally, this book raises more questions than it answers. It is our conviction that architecture needs to continually question its tools, techniques and modes of practice. These questions thicken, expand and open practice so that new expertise and models can emerge, and new aims can be pursued. Our thickened model of drawing particularly calls into question the project-form of architecture: the idea that architectural drawings are primarily anticipatory descriptions pointing away to their future realization. To call this into question is not to dispense with the idea of anticipation or projection. Robin Evans in *The Projective Cast* offers a more complex picture. He concludes his study of architectural drawing with a diagram of architectural practice as a meshwork of 'projective and quasi-projective transactions'.[11] These transactions include the conversion between perspective and orthographic systems and the translation of drawings into buildings, but also the perception of buildings, the perception of drawings as spatial, the conversion of buildings into drawings and even the conversion of perceptions into imaginations. In this model, architecture is projective not because it moves in a straight line from drawing to building, but because it is constantly converting, translating and mediating in all directions.

New permutations of drawing are also new imaginative possibilities for this ongoing mediation. Computation, coding, computer-driven instruments, tools for digital fabrication, artificially intelligent image generators and other technologies can extend architectural imagination, and we have sought to include demonstrations of some of these possibilities. We do not set aside cautions (from Emmons, Scheer and others) that standardized digital techniques and the paradigm of simulation make an indelible mark upon drawing.[12] Yet as our examples show, computation is not necessarily at odds with the body. Computation is now deeply implicated in our lives and habits of living, just as photography and film became in the twentieth century. Our situations are digital and algorithmic as well as organic, social, cultural and so forth. We argue that drawing continues to act as a locus for invention in these situations.

As we signalled in our introduction, we deliberately take a broad view of what constitutes drawing in general and architectural drawing specifically. Historical drawing practices have now been absorbed into a broader culture of imaging and media and have been activated in new ways through paradigms of data and coding. The boundaries between drawing, photography, film, modelling, prototyping and digital simulation are now blurred. Drawing, however, remains important practically: architects and other spatial designers have not stopped drawing. Drawing remains a distinctive architectural activity, even as it continues to take on new forms and methods through its overlaps and encounters with other fields. Given the close links between architectural drawing and the organization of architecture practice, the prompt to draw differently has always also been a prompt to imagine alternative ways of doing architecture. As Catherine Ingraham pointed out over twenty-five years ago: linearity is not just a graphic feature or a stylistic preference in architecture. It is the conceptual model, ideology or imaginary that organizes architectural practice.[13] By drawing in different ways, we do more than communicate with a different voice, we give ourselves the chance to imagine architectural practice differently and position it for new and deeper cross-disciplinary and cross-cultural openings.

There are, naturally, many more ways architectural drawing opens into neighbouring disciplines, practices and cultures. For example, we have not considered the interface between drawing and painting, a practice that has often been more comfortable with its bodily nature and one which many architects have professed as influential. Similarly, while the body often sits uncomfortably in the Western architectural tradition, there are many traditions of spatial thought or creative practice in which the physical presence of the body is fundamental. A glimpse of this, for example, might be seen in *Te Poho o Hine-Moana* (2021), a work by Māori architect and artist Raukura Turei (Ngāi Tai ki Tāmaki, Ngā Rauru Kītahi). Turei has produced a series of works in which she covers canvases with a thick mixture of paint and earth (whenua) and presses through an oil stick, likened to the scale of her finger, to reveal an underpainted surface. In the six joined canvases of *Te Poho o Hine-Moana* the earth is onepū, black sand, and Turei's touch is indexed thousands of time, revealing subtle warmth and shadows that only become visible from a distance.[14] Space and depth accumulate through myriad direct touches, as if the surface is a physical membrane, an interface between one place and another. The mechanisms of architectural drawing (with which Turei is also highly adept) are absent: there are no lines, no projection, and the work is apparently unconcerned with reference. Her work is resolutely full-scale: 'I don't think about them constrained by the stretcher', says Turei, 'I think of them as part of a much bigger expanse, even if they are only a little piece.'[15] In place of line, movement and scale, Turei's paintings access dynamics of covering and revealing, sensuous bodily presence and deep relationships with specific places.

We have also only touched lightly on the way architectural drawing has been enrolled in what anthropologist James Scott famously called the modern 'project of legibility': the use of simplified, schematic views to render things susceptible to knowledge, control and manipulation.[16] These views were at first recognized as abstractions but soon became substitutes for what they depicted and were then projected back onto the world. Surveying – which we argued in Chapter 2 can be seen as full-scale drawing – is 'the crowning artifact of this mighty simplification', according to Scott.[17] Indigenous peoples clearly saw the implications of these drawings across the land: in Aotearoa New Zealand, Māori evicted surveyors, confiscated their equipment and pulled up their pegs as crucial acts of resistance to the conversion of land into a measured commodity.[18] Anthropologist Anna Lowenhaupt Tsing argues that the modern idea of scaling is at the heart of contemporary capitalism and derived from the logic of the sugarcane plantation: 'Projects that could expand through scalability were the poster children of modernization and development. . . . Project advocates thought that they had grasped the world. But [. . .] scalability has left ruins in its wake.'[19] While directly addressing the role that architectural ideation and practice has played in these scenarios falls outside the scope of this book, a richer understanding of the relationship between drawings and bodies opens critical opportunities.

In place of the simplified scale-figure, the standardized anthropometric dummy or the context-independent disembodied eye, the body may be considered, as Elizabeth Grosz poses, as 'a hinge or a threshold'.[20] In full-scale drawing, viewers (or indeed drawers) arrive as themselves, bringing their own complexity to the situation. They may also try on alternative versions of themselves or find they are reflected in unanticipated ways. A consideration of inclusivity is fostered by default, eschewing oversimplification or generalization of the body or too-convenient theoretical models of subject and object. Full-scale drawing questions what happens to drawing and by extension architectural practice, when differences between bodies – variations in capability, sensation, mobility, neurology and experience – come to the surface.

The students we teach now arrive with different experiences, assumptions and aspirations than those with which we entered the field; their careers will involve a breadth of activities unlike those of their professional seniors. They recognize architecture's entanglement with social, cultural, economic, political and institutional structures. They are painfully aware of architecture's complicity in inequality, discrimination and environmental damage. They understand that Western ideas and practices of representation are tightly meshed with colonization and extractive capitalism. We hope that they envision the agency of architectural drawing as not merely reinforcing prevailing patterns and conventions but offering new possibilities.

A drawing, writes Jean-Luc Nancy, is an *opening*: 'a beginning, departure, origin, dispatch, impetus, or sketching out'.[21] Drawing puts architecture in dialogue with other fields: we have focused on relationships with art, but drawing equally opens conversations with geography (through the map), science and organizational studies (through the diagram), music (through notation), ethnography (through field-notes), coding (through the flowchart and digital imaging) and other disciplines. Drawing continues to reconfigure itself. Interrogating assumptions and inventing novel tools are crucial parts of forward-looking architectural practice; these are not only technical matters but also imaginative considerations. By reasserting architectural drawing as speculative, situated and embodied, we claim it as an open-ended mode of inquiry for creative-practice research and critical understandings of contemporary practice.

NOTES

Introduction

1 Of importance here is the work of Paul Emmons and Robin Evans: Paul Emmons, *Drawing Imagining Building: Embodiment in Architectural Design Practices* (Routledge, 2019); Robin Evans, *The Projective Cast: Architecture and Its Three Geometries* (MIT Press, 1995).

2 In this, we draw particularly on accounts by Cornelia H. Butler and Catherine de Zegher, *On Line: Drawing through the Twentieth Century* (Thames & Hudson, 2010); and Rosalind Krauss's seminal *Line as Language: Six Artists Draw* (Princeton University, 1974).

3 For more on this, see Albert Pérez-Gómez and Louise Pelletier, *Architectural Representation and the Perspective Hinge* (MIT Press, 1997); and Dalibor Vesely, *Architecture in the Age of Divided Representation: The Question of Creativity in the Shadow of Production* (MIT Press, 2004).

4 Architectural drawing 'forms questions as much as answers'. Yeoryia Manolopoulou, 'Unformed Drawing: Notes, Sketches, and Diagrams', *The Journal of Architecture* 10, no. 5 (2005): 519, https://doi.org/10.1080/13602360500462323. It is 'sense-making', 'an interactive conversation with the (partial) propositions already put', or 'the medium in a heuristic process' (Simon Twose, Jules Moloney, Anastasia Globa and Lawrence Harvey, 'Drawing the Unfixed', *Interstices: Journal of Architecture and Related Arts*, 21 (2022): 128, https://doi.org/10.24135/ijara.vi.694. See also Peter Downton, *Design Research. Revised Kindle Edition* (Elizabeth James Productions, 2013), Section 6.3.4; Daniel M. Herbert, *Architectural Study Drawings* (Van Nostrand Reinhold, 1993), 116).

5 Paul Emmons describes them as 'akin to divination' (Emmons, *Drawing Imagining Building*, 3); Marco Frascari as a kind of 'savoring' (Marco Frascari, *Eleven Exercises in the Art of Architectural Drawing: Slow Food for the Architect's Imagination* (Routledge, 2011, 69)); Perry Kulper as a 'visual chemical cauldron' (Perry Kulper, 'The Labor of Architectural Drawing', *Figure* 1 [2014]): 10–11).

6 Deanna Petherbridge, *The Primacy of Drawing: Histories and Theories of Practice* (Yale University Press, 2010), 18.

7 Stan Allen, *Practice: Architecture, Technique + Representation* (Routledge, 2009), 48.

8 In the same vein, Robin Evans wrote that drawing is not 'a truck for pushing ideas from place to place but . . . a locale of subterfuges and evasions'. Robin Evans,

Translations from Drawing to Building and Other Essays (Architectural Assocation, 1997), 186.

9 Philosophical approaches to drawing have often distinguished them from writing by emphasizing how they feel their way ahead. For Hélène Cixous this is a kind of ecstasy (Hélène Cixous, 'Without End No State of Drawingness No, Rather: The Executioner's Taking Off', trans. Catherine A.F. MacGillivray, *New Literary History* 24, no. 1 [1993]: 87–103). Jacques Derrida described it in terms of blindness (Jacques Derrida, *Memoirs of the Blind: The Self-Portrait and Other Ruins*, trans. Pascale-Anne Brault and Michael Nas (University of Chicago Press, 1993)). As a 'description without place', it opened to politics for Alain Badiou (Alain Badiou, 'Drawing', *Lacanian Ink* 28 (2006): 42–9). Jean-Luc Nancy describes an 'essential incompleteness . . . non-closure' of drawing (Jean-Luc Nancy, *The Pleasure in Drawing* (Fordham University Press, 2013, 1)).

10 Emmons, *Drawing Imagining Building*, 12.

11 Emmons, *Drawing Imagining Building*, 12.

12 Emmons, *Drawing Imagining Building*, 12, 36.

13 Architect and advocate for drawing by hand Marco Frascari wrote that 'architectural sapience' could only come about processually through 'lingering, savouring and touching': investing time. Marco Frascari, 'Lines as Architectural Thinking', *Architectural Theory Review* 14, no. 3 (2009): 201, https://doi.org/10.1080/13264820903341605.

14 Mario Carpo describes the 1990s and early 2000s as a 'digital turn' in architecture: 'digital tools for design and construction are now unmaking the Albertian, humanistic principles of allographic notation'. Mario Carpo, ed., *The Digital Turn in Architecture 1992–2012* (Wiley, 2013); Mario Carpo, *The Second Digital Turn: Design Beyond Intelligence* (MIT Press, 2017). In 1999, Todd Williams and Billie Tsien wrote that using traditional drawing instruments was a deliberate resistance to the acceleration brought about by the digital: 'As our hands move, we have the time to think and to observe our actions'. Todd Williams and Billie Tsien, 'Slowness', *2G*, no. 9 (1999): 132–6.

15 Allen, *Practice*, xv. Donald Schön argued that practice was characterized by context-sensitive 'reflection-in-action'. Donald Schön, *The Reflective Practitioner: How Professionals Think in Action* (Routledge, 1994), 62.

16 Albena Yaneva, *The Making of a Building: A Pragmatist Approach to Architecture* (Peter Lang, 2009), 6.

17 Instances of measured 1:1 scale include those Le Corbusier made on a large blackboard in his office; the huge section drawing Brunelleschi reputedly made in half a mile of smoothed sand beside the River Arno; or the in situ, taped-out drawings of Flores and Prats Architects for the Sala Beckett. Emmons, *Drawing, Imagining, Building*, 183; Eva Prats, 'To Observe with the Client, to Draw with the Existing: Three Cases of Architecture Dealing with the As-Found' (PhD thesis, RMIT University, 2019), 178.

18 '[W]hat kinds of design might be generated by *starting from* . . . unruly bodies[?]', asks Jos Boys. Jos Boys, ed., *Disability, Space, Architecture: A Reader* (Routledge, 2017), 151. 'How might a collaborative relation of body and environment create the potential for a more non-hierarchical architecture?', ask Lambert and Pham.

Léopold Lambert and Minh-Ha T. Pham, 'Spinoza in a T-Shirt', *The New Inquiry*, 1 July 2015, https://thenewinquiry.com/spinoza-in-a-t-shirt/. See also David Gissen, *The Architecture of Disability: Buildings, Cities, and Landscapes Beyond Access* (University of Minnesota Press, 2023). Elizabeth Grosz asks, 'What does the fact that there are always and irreducibly (at least) two sexes have to do with how we understand and live space?' Elizabeth Grosz, *Architecture from the Outside: Essays on Virtual and Real Space* (MIT Press, 2001, xviii).

19 'Cartesian perspectivalism was thus in league with a scientific world view that . . . saw [the world] as situated in a mathematically regular spatio-temporal order filled with natural objects that could only be observed from without by the dispassionate eye of the neutral researcher'. Martin Jay, 'Scopic Regimes of Modernity', in *Vision and Visuality*, ed. Hal Foster (Bay Press, 1988), 3–28. Edward Said argued that 'descriptive realism' (under which we might group much architectural representation) was a key element in European appropriations of the 'Orient'. Edward Said, *Orientalism: Western Conceptions of the Orient* (Penguin Books, 1995), 87.

20 To draw is to 'produce an image of something by making lines and marks with a pencil, pen, or other implement'. *Oxford English Dictionary*, s.v. 'draw', VI.85.d, https://www.oed.com/dictionary/draw_v?tab=meaning_and_use#6255952.

21 'Let us therefore begin thus: the whole matter of building is composed of lineaments and structure . . . let lineaments be the precise and correct outline, conceived in the mind, made up of lines and angles, and perfected in the learned intellect and imagination'. Leon Battista Alberti, *On the Art of Building in Ten Books*, trans. Joseph Rykwert, Neil Leach and Robert Tavernor (MIT Press, 1988), 7.

22 Alberti, *On the Art of Building*, 7.

23 On the subtle and intricate relationships between drawing and the Renaissance term *disegno* see David Rosand, *Drawing Acts: Studies in Graphic Expression and Representation* (Cambridge University Press, 2016), 24–60.

24 Frascari, 'Lines as Architectural Thinking', 208.

25 'Drawing developed as an index of construction'. Emmons, *Drawing Imagining Building*, 32.

26 'Embedded within these practices is an implicit quasi-theory, or better, a way of thinking invested in a way of acting'. Emmons, *Drawing Imagining Building*, 13.

27 Catherine Ingraham, 'Lines and Linearity: Problems in Architectural Theory', in *Drawing/Building/Text: Essays in Architectural Theory*, ed. Andrea Kahn (Princeton Architectural Press, 1991), 67.

28 Ingraham, 'Lines and Linearity', 67.

29 Catherine Ingraham, *Architecture and the Burdens of Linearity* (Yale University Press, 1998), 68.

30 Ingraham, *Architecture and the Burdens of Linearity*, 68.

31 Tim Ingold, *Lines: A Brief History* (Routledge, 2016), 75.

32 Ross Jenner, 'Thought Out of Bounds: Theory and Practice in Architectural Doctorates', in *Of Other Thoughts: Non-Traditional Ways to the Doctorate*, ed. Anna-Christina Engels-Schwarzpaul and Michael A. Peters (Sense, 2013), 210.

33 Robin Evans, 'Architectural Projection', in *Architecture and Its Image*, ed. Eve Blau, Edward Kaufman and Robin Evans (MIT Press, 1989), 19.

34 Anthony Vidler, 'Diagrams of Diagrams: Architectural Abstraction and Modern Representation', *Representations* 72 (2000): 3, https://doi.org/10.2307/2902906.

35 This is similar to how Evans describes architecture's 'projective cast'. Evans, *Projective Cast*, 366.

36 For a history of descriptive geometry, see Kirsti Andersen, *The Geometry of an Art: The History of the Mathematical Theory of Perspective from Alberti to Monge* (Springer, 2007).

37 For an overview of this, see Carpo, *The Digital Turn in Architecture*.

38 Matthew Allen, *Flowcharting. From Abstractionism to Algorithmics in Art and Architecture* (gta Verlag, 2023), 1. See also Alicia Imperiale's account of artists' adoption of xerography: Alicia Imperiale, 'Machine Consequences', in *Exactitude: On Precision and Play in Contemporary Architecture*, ed. Pari Riahi, Laure Katsaros and Michael T. Davis (University of Massachusetts Press, 2022), 117–33.

39 John May, *Signal. Image. Architecture* (Columbia University Press, 2019), 87.

40 Nancy Foote, 'Drawing the Line', *Artforum* 14, no. 9 (1976): 54–6.

41 As Indra Kagis McEwen points out, his paradigm of order is fundamentally imperial, to the extent that she claims the body – Vitruvian man – is best understood as the emperor's body. Indra Kagis McEwen, *Vitruvius: Writing the Body of Architecture* (MIT Press, 2002), 302.

42 McEwen, *Vitruvius*, 24.

43 McEwen, again, argues that this is fundamentally an imperial order. Indra Kagis McEwen, *All the King's Horses: Vitruvius in an Age of Princes* (MIT Press, 2023).

44 Alberti defined beauty in terms of harmony and proportion; his translators Rykwert, Leach and Tavernor point out that he uses language derived from the Roman writer Cicero's account of order in the human body (*On the Art of Building*, 367). Perspective challenged the idea of fixed proportions. See Evans, *Projective Cast*, 240–71.

45 Diana Agrest, 'Representation as Articulation Between Theory and Practice', in Allen, *Practice*, 165.

46 Justine Clark, 'Smudges, Smears and Adventitious Marks', *Interstices: Journal of Architecture and Related Arts* 4 (1999): 1–8, https://interstices.ac.nz/index.php/Interstices/article/view/568.

47 Christine Macy, 'The Authority of Architectural Drawing', in *Knowledge and/or/of Experience: The Theory of Space in Art and Architecture*, ed. John Macarthur (Institute of Modern Arts, 1993), 146.

48 More precisely, architectural drawings are 'a curious mixture'. Nelson Goodman, *Languages of Art: An Approach to a Theory of Symbols* (Bobbs-Merrill, 1968), 113, 118.

49 'From the concept sketch springs forth the at least metaphoric linearity of architectural production'. Francesca Hughes, *The Architecture of Error: Matter, Measure, and the Misadventures of Precision* (MIT Press, 2014), 35. When asked to provide evidence of his credentials as an architect, Giotto is said to have sketched out a perfect circle, with 'a brush dipped in red, closed his arm to his side, so as to make a sort of compass of it, and then with a twist of his hand drew such a perfect circle that it was a marvel to see'. Giorgio Vasari, *Lives of the Artists*, Volume 1 (Penguin, 1987), 44.

50 'In performing a drawing, the architect feels as well as sees a design, uniting body with understanding'. Emmons, *Drawing Imagining Building*, 12.

51 Ernst Neufert's *Architects' Data* of 1936 was genre-defining. Ernst Neufert and Peter Neufert, *Architects' Data. Third Edition*, ed. Bousmaha Baiche and Nicholas Walliman (Blackwell Science, 2008). Miester examines the ideological freight of Neufert's studies. Anna-Maria Meister, 'Ernst Neufert's "Lebensgestaltungslehre": Formatting Life Beyond the Built', *BJHS Themes* 5 (2020): 167–85, https://doi.org/10.1017/bjt.2020.13. Emmons writes of one anthropometric study at Yale in the 1930s as 'the inscription of ideologies upon the body'. Emmons, *Drawing, Imagining, Building*, 49.

52 According to Gissen, 'experiences of disability are relative to structural inequities', meaning that a narrow functionalist approach, or focus on 'access', is insufficient. Gissen, *The Architecture of Disability*, xiv.

53 Edward S. Casey, *The Fate of Place: A Philosophical History* (University of California Press, 1997), 207.

54 Casey, *Fate of Place*, 208–10.

55 Maurice Merleau-Ponty, *Phenomenology of Perception*, trans. Donald A. Landes (Routledge, 2012), 112.

56 We should reject 'any analysis of bodily space which takes account only of figures and points, since these can neither be conceived nor be without horizons'. Merleau-Ponty, *Phenomenology of Perception*, 115–16.

57 Merleau-Ponty, *Phenomenology of Perception*, 298.

58 Casey, *Fate of Place*, 227.

59 Jin Baek, 'From the "Topos of Nothingness" to the "Space of Transparency": Kitarō Nishida's Notion of Shintai and Its Influence on Art and Architecture', *Philosophy East and West* 58, no. 1 (2008): 83.

60 Baek, 'Kitarō Nishida', 84–5.

61 Elizabeth Grosz, *Space, Time and Perversion: Essays on the Politics of Bodies* (Routledge, 1995), 104.

62 Elizabeth Grosz, *Volatile Bodies: Toward a Corporeal Feminism* (Routledge, 1994), xi.

63 José Gil, 'Paradoxical Body', trans. André Lepecki, *The Drama Review* 50, no. 4 (2006): 28, https://doi.org/10.1162/dram.2006.50.4.21.

64 Bryan Turner, cited in Amelia Jones, *Body Art: Performing the Subject* (University of Minnesota Press, 1998), 12.

65 Carl Douglas, 'Do Not Scale: A Lament for Design Drawing', *TRACEY* 16, no. 1 (2022): 1–13.

66 Javier Fernandez Contreras, *The Miralles Projection: Thinking and Representation in the Architecture of Enric Miralles* (ORO Editions, 2020), 98–9.

67 Emmons points out that the stair also evokes the mental act of walking through a plan (perhaps mirrored by the walking action of measuring distances with the two pointed legs of a pair of dividers). Paul Emmons, 'Drawn to Scale: The Imaginative

Inhabitation of Architectural Drawings', in *From Models to Drawings: Imagination and Representation*, ed. Marco Frascari, Jonathan Hale and Bradley Starkey (Routledge, 2007), 64, 66.

68 Albena Yaneva, 'Scaling Up and Down: Extraction Trials in Architectural Design', *Social Studies of Science* 35 (2005): 868–9. Scale is a means 'to abstract, frame, manipulate, edit, filter, and omit'. Douglas, 'Do Not Scale', 3.

69 Google Earth (2001), based on Keyhole Inc.'s Earth Viewer programme (1999; explicitly drawing on the Eames's film) is 'perhaps the most prominent manifestation and stimulant of this voracious contemporary appetite for views from above'. Mark Dorrian and Frédéric Poussin, *Seeing from Above: The Aerial View in Visual Culture* (I.B. Tauris, 2013), 295. Bill Kilday, *Never Lost Again: The Google Mapping Revolution That Sparked New Industries and Augmented Our Reality* (HarperCollins, 2018), 26.

70 Michael Meredith and Hilary Sample, eds, *An Unfinished Encyclopedia of Scale Figures Without Architecture* (MIT Press, 2019).

71 Michael Young, *Reality Modeled After Images: Architecture and Aesthetics After the Digital Image* (Routledge, 2021), 72.

72 The term derives from the renderings of the École des Beaux-Arts. Young, *Reality Modeled After Images*, 69–98. 'Entourage renders social events and programmatic inhabitation, but it does so in service of evoking attitude, demeanor and interaction'. Young, *Reality Modeled After Images*, 74.

73 Deborah J. Coen, *Climate in Motion: Science, Empire and the Problem of Scale* (University of Chicago Press, 2018), 16.

74 Anna Lowenhaupt Tsing, 'On Non-Scalability: The Living World Is Not Amenable to Precision-Nested Scales', *Common Knowledge* 18, no. 3 (2012): 505, https://doi .org/10.1215/0961754X-1630424.

75 Andrea Kahn and Carol J. Burns, eds, *Site Matters: Strategies for Uncertainty Through Planning and Design* (Routledge, 2021), x.

76 Mark Wigley, 'On Site', *Lotus International* 95 (1997): 119.

77 Wigley, 'On Site', 122–3.

78 Miwon Kwon, *One Place After Another: Site-Specific Art and Locational Identity* (MIT Press, 2002), 1.

79 Kwon, *One Place After Another*, 3. Nick Kaye describes an 'unravelling of notions of original or fixed location' that begins with 'the incursion of performance into visual art and architecture' (Nick Kaye, *Site-Specific Art: Performance, Place and Documentation* (Routledge, 2000), 215; 3). He identifies the site-specific 'with a *working over* of the production, definition and *performance* of "place"'. Kaye, *Site-Specific Art*, 3.

80 Jordan Kauffman, *Drawing on Architecture: The Object of Lines, 1970–1990* (MIT Press, 2018); Zoë Ryan, ed., *As Seen: Exhibitions That Made Architecture and Design History* (The Art Institute of Chicago, 2017); Thordis Arrhenius, Mari Lending, Wallis Miller and Jérémie Michael McGowan, eds, *Exhibiting Architecture: Place and Displacement* (Lars Müller, 2014).

81 Butler and de Zegher, *On Line*, 68.

82 Emmons, *Drawing, Imagining, Building*, 2019.

83 Zeynep Çelik Alexander, *Kinaesthetic Knowing: Aesthetics, Epistemology, Modern Design* (University of Chicago Press, 2017).

84 On the Gilbreths, see Frank Bunker Gilbreth and Robert Thurston Kent, *Motion Study, a Method for Increasing the Efficiency of the Workman* (Van Nostrand Company, 1911); Elspeth H. Brown, *The Corporate Eye: Photography and the Rationalization of American Commercial Culture, 1884–1929* (Johns Hopkins University Press, 2005); Laurel D. Graham, 'Domesticating Efficiency: Lillian Gilbreth's Scientific Management of Homemakers, 1924–1930', *Signs* 24, no. 3 (1999): 633–75; Hyungmin Pai, *The Portfolio and the Diagram: Architecture, Discourse, and Modernity in America* (MIT Press, 2002).

85 Clair Bishop, *Installation Art: A Critical History* (Tate, 2005); Julie H. Reiss, *From Margin to Center: The Spaces of Installation Art* (MIT Press, 2000); Christine Vial Kayser and Sylvie Coëllier, eds, *Installation Art as Experience of Self, in Space and Time* (Vernon Press, 2021).

86 Peter B. Hyland and Tyson E. Lewis, *Studious Drift: Movements and Protocols for a Postdigital Education* (University of Minnesota Press, 2022), 19.

87 Other wide-ranging surveys include: Neil Spiller, *Visionary Architecture: Blueprints of the Modern Imagination* (Thames and Hudson, 2006); Mark Dorrian, Riet Eeckhout and Arnaud Hendrickx, eds, *Drawing Architecture: Conversations on Contemporary Practice* (Lund Humphries, 2022); Laura Allen and Luke Pearson, eds, *Drawing Futures: Speculations in Contemporary Drawing for Art and Architecture* (UCL Press, 2016).

Chapter 1

1 Dan Hoffman, 'Unbuilding and the Recovery of Craft in Architecture: Cranbrook Department of Architecture 1986–1996', *Architectural Design* 93, no. 3 (2023):72, http://dx.doi.org/10.1002/ad.2936.

2 Dan Hoffman, *Architecture Studio: Cranbrook Academy of Art, 1986–1993* (Rizzoli, 1994), 32.

3 Hoffman, *Architecture Studio*, 32.

4 Hoffman, 'Unbuilding', 72. The work was installed at Willis Art Gallery, Detroit, Michigan, in 1989.

5 Hoffman, *Architecture Studio*, 32. A project fifteen years later, in Sydney, Australia, similarly undertook an unbuilding of a suburban house and reassembled it in a gallery space; see Claire Healy and Sean Cordeiro, 'The Cordial Home Project', https://claireandsean.com/large-scale-projects/the-cordial-home-project, accessed 29 November 2024. Five years later, their work, *Not Under My Roof*, dismantled a farmhouse and hung the floor plan on the wall of the Gallery of Modern Art in Brisbane; see Healy and Cordeiro, 'Not Under My Roof', https://claireandsean.com/large-scale-projects/not-under-my-roof, accessed 29 November 2024.

6 Hoffman, 'Unbuilding', 72.

7 Paul Emmons, *Drawing Imagining Building: Embodiment in Architectural Design Practices* (Routledge, 2019), 28–9; Marco Frascari, 'A Reflection on Paper and Its

Virtues Within the Material and Invisible Factures of Architecture', in *From Models to Drawings: Imagination and Representation in Architecture*, ed. Marco Frascari, Jonathan A. Hale and Bradley Starkey (Routledge, 2007), 24–6.

8 Mario Carpo, *The Digital Turn in Architecture 1992–2012* (Wiley, 2013), 8.

9 Kate Ganz, *Drawing Then: Innovation and Influence in American Drawings of the Sixties* (Dominique Lévy, 2016), 9.

10 Rosalind Krauss, *Line as Language: Six Artists Draw* (Princeton University, 1974), 7.

11 Dorothea Rockburne, 'Notes to Myself on Drawing', *Flash Art* April (1974): 66.

12 Krauss, *Line as Language*, 23.

13 Naomi Spector, 'Dorothea Rockburne', in *Mel Bochner, Barry Le Va, Dorothea Rockburne, Richard Tuttle: Exhibition, Jan 8–Feb 16, 1975* (Contemporary Arts Center, 1975), 31.

14 Anna Lovatt, 'Dorothea Rockburne: Intersection', *October* 122 (2007): 47, https://doi.org/10.1162/octo.2007.122.1.31.

15 Alain Seban, 'Foreword', in *Sol LeWitt*, ed. Beatrice Gross (Centre Pompidou-Metz; M-Museum; Terra Foundation for American Art, 2012), 7.

16 LeWitt may have observed this separation working as a graphic designer in I. M. Pei's architectural practice in the mid-1950s. Gross, 'Order and Disorder', in Gross, *Sol LeWitt*, 21.

17 Anna Lovatt, 'Ideas in Transmission: LeWitt's Wall Drawings and the Question of the Medium', *Tate Papers* 14 (1 October 2010), https://www.tate.org.uk/research/tate-papers/14/ideas-in-transmission-lewitt-wall-drawings-and-the-question-of-medium.

18 Fred Sandback, *Fred Sandback: Vertical Constructions*, ed. David Gray and Lucas Zwirner (David Zwirner Books, 2017), 42.

19 Reinhard Spieler, 'Foreword', in Fred Sandback, *Fred Sandback: Räume zeichnen / Drawing Spaces; Arbeiten auf Papier / Works on Paper* (Kerber, 2011), 7.

20 James Lawrence, in Sandback, *Fred Sandback: Räume zeichnen*, 22.

21 Krauss, *Line as Language*, 32.

22 Cited in Gross, 'Order and Disorder', 11. Also, Sol LeWitt stated, 'Obviously a drawing of a person is not a real person, but a drawing of a line is a real line.' Andrew Wilson, 'Sol LeWitt Interviewed', *Art Monthly* March (1993): 3. Similarly, Sandback wrote: 'There isn't an idea which transcends the actuality of the pieces. The actuality is the idea.' Sandback, *Fred Sandback: Räume zeichnen*, 88.

23 Rosalind Krauss, 'This New Art: To Draw in Space', in *The Originality of the Avant-Garde and Other Modernist Myths* (MIT Press, 1985), 124.

24 Catherine de Zegher, 'A Century Under the Sign of Lines: Drawing and Its Extension (1910–2010)', in *On Line: Drawing Through the Twentieth Century*, ed. Cornelia H. Butler and Catherine de Zegher (Museum of Modern Art, 2011), 68.

25 Claire Bishop, *Installation Art: A Critical History* (Tate Publishing, 2005), 6, https://archive.org/details/installationartc0000bish_v8d6.

26 Lissitzky wrote: 'We saw that the surface of the Proun ceases to be a picture and turns into a structure round which we must circle, looking at it from all sides, peering down from above, investigating from below.' El Lissitzky, 'Proun – Not World

Visions, but World Reality' (1920), in *El Lissitzky: Life, Letters, Text*, ed. Sophie Lissitzky-Küppers (Thames and Hudson, 1968), 343.

27 El Lissitzky, 'The Film of El's Life' (1928), in Lissitzky-Küppers, *El Lissitzky*, 325.

28 Christina Lodder, 'Seeing Red: Lissitzky's Abstract Cabinet and the Ideology of Display', *Umění Art* 6 (1999): 518–19. De Zegher writes: 'From the beginning of the exploration of line in the twentieth century, hierarchies of drawing, painting, and sculpture, of architecture, typography, and environment were broken down.' De Zegher, 'A Century Under the Sign of Lines', in Butler and de Zegher, *On Line*, 48. Other important crossovers between drawing and installation discussed by de Zegher include spatial constructions by Alexandr Rodchenko, Alexander Calder, Marcel Duchamp, Lygia Clark and Eva Hesse.

29 Bishop, *Installation Art*, 6.

30 Bishop, *Installation Art*, 11.

31 Hans Ulrich Obrist, 'Installations Are the Answer, What Is the Question?' *Oxford Art Journal* 24, no. 2 (2001): 93, https://doi.org/10.1093/oxartj/24.2.93.

32 Fred Sandback, 'Untitled', in *Fred Sandback* (Kunstmuseum Liechtenstein; Fruitmarket Gallery; Neue Galerie am Landesmuseum Joanneum; CAPC musée d'art contemporain; Hatje Cantz, 2005), 152, https://www.hatjecantz.com/products/15507-fred-sandback?srsltid=AfmBOordBmQczPSf5m2POqAHRWpJ7IJdDDZBDPzatTdsJg_nKNcTU0KW; cited in *Fred Sandback: Vertical Constructions*, ed. David Gray and Lucas Zwirner (David Zwirner Books, 2017), 43.

33 Miwon Kwon writes, in reference to this time: 'And the space of art was no longer perceived as a blank slate, a tabula rasa, but a real place. . . . Site-specific work in its earliest formation, then, focused on establishing an inextricable, indivisible relationship between the work and its site, and demanded the physical of the viewer for the work's completion.' Miwon Kwon, *One Place After Another: Site-Specific Art and Locational Identity* (MIT Press, 2002), 11–12.

34 Mel Bochner, *Measurement Room: No Vantage Point* (Dia Beacon, New York, 1969/2019), https://www.diaart.org/program/past-programs/mel-bochner-exhibition/year/2019.

35 Katharine Stout, *Contemporary Drawing from the 1960s to Now* (Tate Publishing, 2014), 23.

36 Mark Wigley, *Cutting Matta-Clark: The Anarchitecture Investigation* (Lars Müller Publishers, 2018), 169.

37 Wigley, *Cutting Matta-Clark*, 177.

38 Alan Moore, 'Gordon Matta-Clark: John Gibson Gallery', *Artforum* (December 1974): 81; cited in Wigley, *Cutting Matta-Clark*, 188.

39 Wigley, *Cutting Matta-Clark*, 206.

40 Francesca Hughes, *The Architecture of Error: Matter, Measure, and the Misadventures of Precision* (MIT Press, 2014), 142.

41 Hughes, *Architecture of Error*, 142.

42 Matta-Clark explicitly saw his practice as drawing: 'Why hang things on a wall when the wall itself is so much more a challenging medium? . . . A simple cut or series of cuts, act as a powerful drawing device able to redefine spatial situations and structural components. What is invisibly at play behind a wall or floor, once exposed, becomes an active participant in a spatial drawing of the building's inner life.' Gordon Matta-Clark, 'Interview with Gordon Matta-Clark, Antwerp, September 1977', reprinted from the catalogue *Gordon Matta-Clark* (International Cultureel Centrum, Antwerp, 1977), in *Gordon Matta-Clark 1943–1978*, ed. Corinne Diserens; survey by Thomas Crow; essays by Judith Russi Kirshner and Christian Kravagna (Phaidon Press, 2003), 188, https://www.si.edu/object/siris_sil_770722.

43 Since completed buildings cannot be exhibited, drawings and models present architecture, taken to be architecture's representation; Jean-Louis Cohen writes: 'It is not the *ouvrage*, or the completed building, that is physically present in the exhibition, unless in the case of a prototype or a modular building; it is the *œuvre*, i.e. the project, the intellectual work crystallized in sketches, scale drawings, texts, and of course, models.' Jean-Louis Cohen, 'Models and the Exhibition of Architecture', in *The Art of Architecture Exhibitions*, ed. Kristin Feireiss and Jean-Louis Cohen (Architecture Institute, Rotterdam, 2001), 30.

44 Zoë Ryan, 'Taking Positions: An Incomplete History of Architecture and Design Exhibitions', in *As Seen: Exhibitions That Made Architecture and Design History*, ed. Zoë Ryan (Art Institute of Chicago, 2017), 13.

45 For example, the Museum of Modern Art in New York mounted *Visionary Architecture* (29 September–4 December 1960), https://www.moma.org/calendar/exhibitions/2554 and *Architectural Fantasies: Drawings from the Museum Collection* (27 July 1967–12 February 1968), https://www.moma.org/calendar/exhibitions/1776. Jordan Kauffman, *Drawing on Architecture: The Object of Lines, 1970–1990* (MIT Press, 2018), 10.

46 The show consisted of 108 drawings, 22 models and 9 photographs. Kauffman, *Drawing on Architecture*, 143.

47 A total of 25,000 people were estimated to have seen the show, and it instigated four subsequent shows between 1977 and 1983. Kauffman, *Drawing on Architecture*, 140, 169.

48 Kauffman, *Drawing on Architecture*, 2. Contemporaneously, significant collections of architectural drawings were established; notable are those by Phyllis Lambert, which became the foundation of the Canadian Centre for Architecture's archive and by Barbara Pine and Judith York Newman.

49 Kauffman, *Drawing on Architecture*, 163.

50 Kauffman highlights James Stirling's model for the Cambridge University History Faculty Building as a notable exception: it was suspended from the ceiling in a clear case so it could be viewed from underneath. Kauffman, *Drawing on Architecture*, 147.

51 Speculative drawings which were not intended to bring about a built outcome had a long history, including, for example, works by Étienne-Louis Boullée and Giovanni Battista Piranesi in the eighteenth century, Barthélemy Prosper Enfantin in the nineteenth century, El Lissitzky, Yakov Chernikhov and Theo Van Doesburg in the early twentieth century, and Archigram in the 1960s.

52 Kauffman, *Drawing on Architecture*, 2.

53 Thordis Arrhenius, 'Discourse', in *Exhibiting Architecture: Place and Displacement*, ed. Thordis Arrhenius, Mari Lending, Wallis Miller and Jérémie Michael McGowan (Lars Müller, 2014), 17.

54 Kauffman, *Drawing on Architecture*, 174.

55 Kauffman, *Drawing on Architecture*, 174.

56 Raimund Abraham, *Ungebaut/Unbuilt* (Forum AR/GE Kunst, 1986), 114.

57 Jordan Kauffman, 'Model Values', in *Valuing Architecture: Heritage and the Economics of Culture*, ed. Ashley Paine, Susan Holden and John Macarthur (Valiz, 2020), 71–2.

58 Excerpt from letter sent by the Institute for Architecture and Urban Studies to solicit models for the exhibition. Kenneth Frampton and Sylvia Kolbowski, eds, *Idea as Model* (Rizzoli International Publications, 1981), 3.

59 Rafael Moneo, 'A Study of a Wall with Windows', in Frampton and Kolbowski, *Idea as Model*, 56–7; Stefaan Vervoort, 'Scale Models and Postmodernism: Revisiting Idea as Model (1976–1981)', *Architectural Theory Review*, 24, no. 3 (2020): 225.

60 Rafael Moneo, 'A Study of a Wall with Windows', 56.

61 Charles Moore and Richard Oliver, 'A House Near New York', in Frampton and Kolbowski, *Idea as Model*, 58.

62 Bernhard Leitner, 'Four Spatial Transformations of a Grid', in Frampton and Kolbowski, *Idea as Model*, 50–1.

63 Richard Pommer, 'Postscript to a Post-Mortem', in Frampton and Kolbowksi, *Idea as Model*, 15; Richard Pommer, 'The Idea of "Idea as Model"', in Frampton and Kolbowski, *Idea as Model*, 9.

64 Christian Hubert, 'The Ruins of Representation', in Frampton and Kolbowksi, *Idea as Model*, 17.

65 Frampton and Kolbowski, *Idea as Model*, 6.

66 Pamela M. Lee, *Object to Be Destroyed: The Work of Gordon Matta-Clark* (MIT Press, 2000), 116.

67 Christian Hubert, '"The Ruins of Representation" Revisited', *OASE* 84 (2011): 19.

68 For more on full-scale models using studio-museums as case studies, see Marian Macken, 'Documenting Interiority / Inhabiting Duration', in *The Interior Architecture Theory Reader*, ed. Gregory Marinic (Routledge, 2018), 310–17.

69 Jonathan Hill, 'Weathering the Barcelona Pavilion', *Journal of Architecture* 7, no. 4 (2002): 321, https://doi.org/10.1080/13602360320000408 20.

70 Installed as part of Ars Electronica Festival; documentation was later included in the exhibition *Light Construction*, curated by Terence Riley at the Museum of Modern Art, New York (21 September 1995–2 January 1996), https://www.moma.org/calendar/exhibitions/469.

71 MeGo, 'Floor Plan, Melissa Gould', http://www.megophone.com/floorplan.html, accessed 16 September 2024.

72 Victoria Milne, 'Bringing the Past to Light', *Metropolis Magazine* January/February (1994): 33.

73 Melissa Gould, quoted in Terrence Riley, *Light Construction* (Museum of Modern Art, 1995), 136.

74 Bishop, *Installation Art*, 130.

75 Hoffman, *Architecture Studio*, 187.

76 Hoffman, *Architecture Studio*, 189.

77 Kent Kleinman, 'Not-P', in *Gravity: Pamphlet Architecture* 25, ed. James Cathcart, Frank Fantauzzi and Terence Van Elslander (Princeton Architectural Press, 2003), 12.

78 Kleinman, 'Not-P', 12.

79 In confronting Western architectural assumptions, '[d]istinctions either evaporate or at the very least need to be refracted through an indigenous lens . . . making them not only inherently untenable but likely to be subverted if not dismantled'. Hirini Mutunga, 'A Discourse on the Nature of Indigenous Architecture', in *The Handbook of Contemporary Indigenous Architecture*, ed. Elizabeth Grant, Kelly Greenop, Albert L. Refiti and Daniel J. Glenn (Springer, 2018), 325.

Chapter 2

1 Euclid, *Euclid's Elements of Geometry*, ed. Richard Fitzpatrick (Richard Fitzpatrick, 2008), 6. Paul Emmons writes that prior to the Renaissance, Euclid rather than Vitruvius was of critical importance to architecture. Paul Emmons, *Drawing Imagining Building: Embodiment in Architectural Design Practices* (Routledge, 2019), 24. Robin Evans describes the grounding role Euclidean geometry plays for Renaissance architects (Robin Evans, *The Projective Cast: Architecture and Its Three Geometries* (MIT Press, 1995), xxxii). It is important to note that what seems like purely mathematical statements like this have a 'ritual origin' in the construction of 'sacred figures'. Abraham Seidenberg, 'The Ritual Origin of Geometry', *Archive for History of Exact Sciences* 1, no. 5 (1962): 523.

2 Robert Hooke, *Micrographia: Or Some Physiological Descriptions of Minute Bodies Made by Magnifying Glasses with Observations and Inquiries Thereupon* (Jo. Martyn and Ja. Allestry, 1665), 3. This passage is discussed by Cynthia Sundberg Wall (Cynthia Sundberg Wall, *The Prose of Things: Transformations of Description in the Eighteenth Century* (University of Chicago Press, 2014), 72–6); Francesca Hughes, *The Architecture of Error: Matter, Measure, and the Misadventures of Precision* (MIT Press, 2014), 17–18; and Paul Emmons, 'Drawn to Scale: The Imaginative Inhabitation of Architectural Drawings', in *From Models to Drawings: Imagination and Representation in Architecture*, ed. Marco Frascari, Jonathan Hale and Bradley Starkey (Routledge, 2007), 69–70. For Hooke's life, see Lisa Jardine, *The Curious Life of Robert Hooke: The Man Who Measured London* (Harper Perennial, 2003); on Hooke's surveying and architectural work, see notes 34 and 37 below.

3 Hooke, *Micrographia*, 1.

4 Hooke *Micrographia*, 2–3.

5 René Descartes (1596–1650) established his famous split in his Second Meditation. René Descartes, *The Philosophical Writings of Descartes, Vol. II*, trans. John Cottingham, Robert Stoothoff and Dugald Murdoch (Cambridge University Press, 1984), 16–23. On the specific implications for European understandings of space, see Edward S. Casey, *The Fate of Place: A Philosophical History* (University of California Press, 1997), 150–60.

6 Francesca Hughes, *The Architecture of Error: Matter, Measure, and the Misadventures of Precision* (MIT Press, 2014), 39, 4.

7 Emmons, *Drawing Imagining Building*, 217. David Ross Scheer describes this as 'simulation' rather than drawing. David Ross Scheer, *The Death of Drawing: Architecture in the Age of Simulation* (Routledge, 2014), 15.

8 See Mario Carpo, ed., *The Digital Turn in Architecture 1992–2012* (John Wiley, 2012) and Mario Carpo, *The Second Digital Turn: Design Beyond Intelligence* (MIT Press, 2017). Lefaivre and Tzonis argued that geometry, optimization and verification provided 'the pre-existing conceptual pieces' of mechanization and ultimately computation. Liane Lefaivre and Alexander Tzonis, 'Die Maschine Im Architektonischen Denken: The Machine in Architectural Thinking', *Daidalos* 18 (1985): 16–26. See also Alexander Galloway, *Uncomputable: Play and Politics in the Long Digital Age* (Verso Books, 2021).

9 Leon Battista Alberti, *On Painting and On Sculpture: The Latin Texts of De Pictura and De Statua*, ed. Cecil Grayson (Phaidon, 1972), 55.

10 For a technical account of this process, see Kristi Andersen, *The Geometry of an Art: The History of the Mathematical Theory of Perspective from Alberti to Monge* (Springer, 2007). For a philosophical one, see Alberto Peréz-Gomez and Louise Pelletier, *Architectural Representation and the Perspective Hinge* (MIT Press, 1997). The literature on the origins of perspective is intricate, but Erwin Panofsky's *Perspective as Symbolic Form* (Zone Books, 1991), https://www.zonebooks.org/books/87-perspective-as-symbolic-form; and Hubert Damisch's *The Origin of Perspective* (MIT Press, 1994) are well-established accounts.

11 Andersen, *Geometry of an Art*, 208. Albrecht Dürer, *The Painter's Manual: A Manual of Measurement of Lines, Areas, and Solids by Means of Compass and Ruler Assembled by Albrecht Dürer for the Use of All Lovers of Art with Appropriate Illustrations Arranged to Be Printed in the Year MDXXV*, trans. Walter L. Strauss (Abaris Books, 1977 [1525]). The image is often discussed, including by Evans, *The Projective Cast*, 127–9.

12 Anderson, *Geometry of an Art*, 208.

13 *Temple Island* was first exhibited at the Architectural Association in 1987 and evolved into an open-ended, decades-long investigation (*The Journey*) consisting now of an extensive archive of drawings, diagrams, renderings, paintings, notes, collages and models. Michael Webb, Michael Sorkin and Lebbeus Woods, *Temple Island: A Study* (Architectural Association, 1987); Ashley Simone, ed., *Michael Webb: Two Journeys* (Lars Müller Publishers, 2018).

14 According to Webb, the image 'unabashedly [tries] to rip off Claude Lorrain'. Webb, *Temple Island*, 1. In the Claude Glass, '[t]he reduction of the sensible dimensions is compensated for by the acquisition of intelligible dimensions'. Arnaud Maillet, *The Claude Glass: Use and Meaning of the Black Mirror in Western Art* (Zone Books, 2009), 86.

15 Michael Sorkin, 'Michael Webb, Temple Island: AA Exhibition Gallery, 1–30 May 1987', *AA Files* 15 (1987): 67.

16 Albrecht Dürer, *Vier Bücher von Menschlicher Proportion*, ed. Berthold Hinz (Akademie Verlag, 2011).

17 Piero described his technique in *De Prospectiva Pingendi*, written some time before 1482. Piero della Francesca, *De Prospectiva Pingendi*, ed. Chiara Gizzi (Fondazione Università Ca' Foscari, 2016), http://doi.org/10.14277/978-88-6969-099-0. Various handwritten manuscripts can be compared at https://exhibits.museogalileo.it/deprospectivapingendi/.

18 Evans, *Projective Cast*, 151.

19 '[T]he sublime culmination of a magical (mathematical) network of connections extending through the universe'. Péréz-Gomez and Pelletier, *Architectural Representation*, 38.

20 Evans, *Projective Cast*, 154. Dürer made similar sectional samples of the body. Dürer, *Vier Bücher*, 326.

21 Alberti, *On Painting*, 130–1. Evans notes a second method described by Alberti, in which a statue is encased in a cylinder of wax which can then be probed for depth at regular intervals. Evans, *Projective Cast*, 156.

22 Alberti, *On Painting*, 135. Alberti too described his method as a process of cutting sectional profiles (133).

23 Contemporary photogrammetric methods build on these techniques. In the nineteenth century, French photographer and artist François Willème constructed a working system of scanning the human body using multiple cameras. Alexander Galloway, *Uncomputable: Play and Politics in the Long Digital Age* (Verso Books, 2021), 31.

24 Michael Young, *Reality Modeled After Images: Architecture and Aesthetics after the Digital Image* (Routledge, 2021), 51.

25 Carl Douglas, 'Point/Cloud: Diffusive Spatial Imaginaries', *idea journal* 20, no. 1 (2023): 26–43, https://doi.org/10.37113/ij.v20i01.496. For a comparison with photogrammetry, see Young, *Reality Modeled after Images*, 55–6.

26 Russ Flatt, 'Te Ahua, Te Wa, Te Atea', https://www.russflatt.com/te-ahua-te-wa-te-atea, accessed 22 September 2024.

27 Young, *Reality Modelled After Images*, 57.

28 Edward Husserl, *The Crisis of European Sciences and Transcendental Phenomenology: An Introduction to Phenomenological Philosophy*, trans. David Carr (Northwestern University, 1970), 48.

29 Leon Battista Alberti, *Descriptio Urbis Romae*, ed. Jean-Yves Boriaud and Francesco Furlan (Casa Editrice Leo C. Olschki, 2005). See also Kim Williams, Lionel March and Stephen R. Wassell, eds, *The Mathematical Works of Leon Battista Alberti* (Birkhäuser, 2010); and Jessica Maier, *Rome Measured and Imagined: Early Modern Maps of the Eternal City* (University of Chicago Press, 2015), 26.

30 Indra Kagis McEwen, *All the King's Horses: Vitruvius in an Age of Princes* (MIT Press, 2023), 35.

31 Maier, *Rome Measured*, 31.

32 Leon Battista Alberti, *On the Art of Building in Ten Books*, trans. Joseph Rykwert, Neil Leach and Robert Tavernor (MIT Press, 1988), 336.

33 Alberti, *On the Art of Building*, 336.

34 On Hooke's surveying, see M. A. R. Cooper, 'Robert Hooke's Work as Surveyor for the City of London in the Aftermath of the Great Fire. Part One: Robert Hooke's First Surveys for the City of London', *Notes and Records of the Royal Society of London* 51, no. 2 (1997): 161–74.

35 Nuala Gregory, 'Encountering Drawing', *Educational Philosophy and Theory* 49, no. 13 (2017): 1240.

36 For this reason, surveying has proved a crucial instrument of colonization and state power. 'Precisely because they are abstract units detached from any ecological or topographical reality, they resemble a kind of currency which is endlessly amenable to aggregation and fragmentation. This feature of the grid plan suits equally the surveyor, the planner, and the real-estate speculator.' James Scott, *Seeing like a State: How Certain Schemes to Improve the Human Condition Have Failed* (Yale University Press, 1999), 58; see also Giselle Byrnes, *Boundary Markers: Land Surveying and the Colonisation of New Zealand* (Bridget Williams Books, 2001).

37 Until 1763, the monument was the highest point in London. For more on Hooke's architectural practice, see Lisa Jardine, *The Curious Life of Robert Hooke: The Man Who Measured London* (Harper Perennial, 2003); and Matthew F. Walker, 'The Limits of Collaboration: Robert Hooke, Christopher Wren and the Designing of the Monument to the Great Fire of London', *Notes and Records of the Royal Society of London* 65, no. 2 (2011): 121–43.

38 Traffic vibrations prevented the telescope from being used as intended, although Hooke found other scientific uses for the tower. Maria Zack, 'Robert Hooke's Fire Monument: Architecture as a Scientific Instrument', in *Architecture and Mathematics from Antiquity to the Future: Vol. 2: The 1500s to the Future*, ed. Kim Williams and Michael J. Ostwald (Springer, 2015), 265.

39 'Drawing was an index of construction.' Emmons, *Drawing Imagining Building*, 13.

40 Ivan Edward Sutherland, 'Sketchpad, A Man-Machine Graphical Communication System' (PhD diss., Massachusetts Institute of Technology, 1963). Sutherland's famous demonstration can be viewed here: 'Ivan Sutherland Sketchpad Demo 1963', YouTube, https://youtu.be/6orsmFndx_o, accessed 24 September 2024.

41 Hughes, *Architecture of Error*, 20–1.

42 Sutherland, 'Sketchpad', 8. These conversational drawings, however, are no longer single configurations: 'there is no state of the system that can be called "drawing". . . . With a computer sketch . . . any line segment is straight and can be relocated by moving one or both of its end points' (102).

43 Young, *Reality Modeled After Images*, 50.

44 Alain Badiou, 'Drawing', *Lacanian Ink* 28 (2006): 42–9.

45 Richard Tuttle, cited in *On Line: Drawing Through the Twentieth Century*, ed. Cornelia H. Butler and M. Catherine de Zegher (Museum of Modern Art, 2011), 119.

46 Emmons, *Drawing Imagining Building*, 99.

47 Lewis Carroll, *Sylvie and Bruno Concluded* (Macmillan, 1894), 169.

48 Jorge Luis Borges, *Collected Fictions* (Penguin Books, 1999), 325.

49 *Oxford English Dictionary*, s.v. 'zoom (int), Etymology', March 2024, https://www.oed.com/dictionary/zoom_int?tab=meaning_and_use#13505189.

Chapter 3

1 Catherine de Zegher, 'A Century Under the Sign of Lines: Drawing and Its Extension (1910–2010)', in *On Line: Drawing Through the Twentieth Century*, ed. Cornelia H. Butler and M. Catherine de Zegher (Museum of Modern Art, 2011), 99.

2 Jim Allen quoted in Robert Leonard, *Art New Zealand* 95 (2000), https://robertleonard.org/jim-allen-contact/. The work was restaged at the Govett-Brewster Art Gallery, New Zealand (2010) and at Artspace, Auckland (2011) with live feed to a lecture theatre in the Faculty of Architecture and Design, Te Herenga Waka Victoria University of Wellington. Jim Allen, *The Skin of Years: Interviews with Jim Allen by Phil Dadson and Tony Green* (Clouds & Michael Lett, 2014), 277–8.

3 Allen's use of the body as a drawing instrument had been preceded by earlier similar work such as Robert Rauschenberg and Susan Weil's bodily blueprints (1949–51), Yves Klein's *Anthropométries* (1958) and Jasper John's *Studies for Skin* (early 1960s); and the lineage continued long after, for example, Janine Antoni's *Loving Care* (1993).

4 'Carolee Schneemann to Daryl Chin: Regarding *Up To and Including Her Limits*, 28 May 1975', in Carolee Schneemann Papers 1959–94, The Getty Research Institute, Los Angeles, cited in Laurence Schmidlin, 'Drawing as a Tool for Mapping the Body in Space', in *Body, Space, and Place in Collective and Collaborative Drawing; Drawing Conversations II*, ed. Jill Journeaux, Helen Gørrill and Sara Reed (Cambridge Scholars Publishing, 2020), 43.

5 Butler and de Zegher, *On Line*, 100.

6 YouTube, 'Rachael Lincoln performs Trisha Brown's "Man Walking Down the Side of a Building"', https://www.youtube.com/watch?v=TbpBHTPODB4, accessed 11 November 2024. Brown followed this work with *Woman Walking Down a Ladder* (1973).

7 Maggie Nelson, *Like Love: Essays and Conversations* (Penguin Random House, 2024), 109.

8 Robin Evans, *Translations from Drawing to Building and Other Essays* (AA Publications, 1997), 224.

9 Zeynep Çelik Alexander, *Kinaesthetic Knowing: Aesthetics, Epistemology, Modern Design* (University of Chicago Press, 2017), 141.

10 Obrist, quoted in Alexander, *Kinaesthetic Knowing*, 160.

11 Paul Klee, *Notebooks. Volume 1. The Thinking Eye*, ed. Jürg Spiller, trans. Ralph Manheim (Lund Humphries, 1961), 105.

12 Klee, *Notebooks*, 109.

13 Giuliana Bruno writes that moving images were taken to be 'the very epitome of modernity'. Giuliana Bruno, *Atlas of Emotion: Journeys in Art, Architecture, and Film* (Verso, 2018), 16.

14 Frank Bunker Gilbreth and Robert Thurston Kent, *Motion Study: A Method for Increasing the Efficiency of the Workman* (D. Van Nostrand Company, 1911).

15 Elspeth H. Brown, *The Corporate Eye: Photography and the Rationalization of American Commercial Culture, 1884–1929* (Johns Hopkins University Press, 2005), 86.

16 Brown, *Corporate Eye*, 93.

17 Elizabeth Diller and Ricard Scofidio write: 'Time-motion studies which had been developed to dissect every action of the factory laborer with the intention of designing ideal shapes of movement and, ultimately, the ideal laborer, were imported into the home to scrutinize every movement exerted in housekeeping in order to produce the ideal housewife.' Elizabeth Diller and Ricardo Scofidio, *Flesh: Architectural Probes* (Princeton Architectural Press, 1994), 41.

18 Laurel D. Graham, 'Domesticating Efficiency: Lillian Gilbreth's Scientific Management of Homemakers, 1924–1930', *Signs* 24, no. 3 (1999): 652–3, https://doi.org/10.1086/495368.

19 Ben Cosgrove, 'Behind the Picture: Picasso Draws with Light', *Life*, https://www.life.com/arts-entertainment/behind-the-picture-picasso-draws-with-light/, accessed 1 October 2024. This was a development of Man Ray's 'light writing' technique used in images such as *Space Writing (Self-Portrait)* (1937).

20 Frederick Kiesler, 'Pseudo-Functionalism in Modern Architecture', *Partisan Review* 16, no. 7 (1949): 734.

21 Kiesler, 'Pseudo-Functionalism', 734.

22 This recalls Gaston Bachelard's idea of a 'garment-house' shaped by and perfectly fitted to the body. Gaston Bachelard, *The Poetics of Space* (Beacon Press, 1994), 101.

23 Kielser, 'Pseudo-Functionalism', 740.

24 Edwin Heathcote, *Imre Makovecz: The Wings of the Soul* (Academy Editions, 1997), 28.

25 Heathcote, *Imre Makovecz*, 120.

26 Yeoryia Manolopoulou, *Architectures of Chance* (Routledge, 2013), 117.

27 Frank Gehry, quoted in Calvin Tomkins, 'The Maverick', *New Yorker* 7 July (1997): 43.

28 Jeremy Till, *Architecture Depends* (MIT Press, 2009), 75–7; 94.

29 On the tenses and cases of architectural drawing, see Robin Evans, 'Architectural Projection', in *Architecture and Its Image: Four Centuries of Architectural Representation: Works from the Collection of the Canadian Centre for Architecture*, ed. Eve Blau and Edward Kaufman (Canadian Centre for Architecture; MIT Press, 1989), 19–35; Robin Evans, 'Translations from Drawing to Building', in Evans, *Translations from Drawing to Building*, 153–93; Alberto Pérez-Gómez and Louise Pelletier, *Architectural Representation and the Perspective Hinge* (MIT Press, 1997); and Alberto Pérez-Gómez, 'The Revelation of Order: Perspective and Architectural Representation', in *This Is Not Architecture: Media Constructions*, ed. Kester Rattenbury (Routledge, 2002), 3–25.

30 David Leatherbarrow, 'Showing What Otherwise Hides Itself: On Architectural Representation', *Harvard Design Magazine* Fall (1998): 53.

31 *The Stage of Drawing: Gesture and Act, Selected from the Tate Collection by Avis Newman, Curated by M. Catherine de Zegher* (Tate Publishing; The Drawing Center, 2003), 95. Bryson, in referring to the difference between drawing and painting, writes that the drawn line 'in a sense always exists in the present tense, in the time of its own unfolding, the ongoing time of a present that constantly presses forward'. Norman Bryson, 'A Walk for Walk's Sake', in *The Stage of Drawing*, 149–50.

32 Gustau Gili Galfetti, ed., *Allan Wexler* (GG Portfolio, 1998), 21.

33 Anthony Vidler, *The Architectural Uncanny: Essays in the Modern Unhomely* (MIT Press, 1992), 210.

34 Michael Young, *Reality Modeled After Images: Architecture and Aesthetics After the Digital Image* (Routledge, 2022), 118.

35 Dan Hoffman, *Architecture Studio: Cranbrook Academy of Art, 1986–1993* (Rizzoli, 1994), 169.

36 Hoffman, *Architecture Studio*, 110.

37 Brown, *Corporate Eye*, 70.

38 Francesca Hughes, *The Architecture of Error: Matter, Measure, and the Misadventures of Precision* (MIT Press, 2014), 5.

39 John James, *The Contractors of Chartres*, Vol. 2 (London: Mandorla Publications, 1979), 544, cited in Peter Wood, 'Drawing the Line: A Working Epistemology for the Study of Architectural Drawing' (PhD diss., University of Auckland, 2002), 348.

40 Hoffman, *Architecture Studio*, 13.

41 Hoffman, *Architecture Studio*, 14.

42 Hoffman, *Architecture Studio*, 14.

43 Matthew Allen, *Flowcharting: From Abstractionism to Algorithmics in Art and Architecture* (gta Verlag, 2023), 125.

44 Gaia Piccarolo, 'Textile Architecture', *Lotus International* 176 (2024): 48–53. Work from the project was exhibited in installation form at the 2018 Venice Biennale, https://www.dipdiitextiles.org.

45 Piccarolo, 'Textile Architecture', 48.

46 Research assistance by Jacky Ziyi Zheng, Stephen Njoto and Maito Akiyama.

47 For more on process, see Aaron Paterson, Sarosh Mulla and Marian Macken, 'Drawing the Room | Drawing within the Room', *Drawing: Research, Theory, Practice* 5, no. 2 (2020): 271, https://doi.org/10.1386/drtp_00036_1.

48 Included in the 2019 New Zealand Institute of Architects' Festival of Architecture; an edited video of the work was included in Ars Electronica Festival for Art, Technology and Society: Garden Aotearoa New Zealand (2020), https://www.ars.nz/drawing-the-room/.

49 Karamia Müller, 'Relational Drawing as Agency: Negotiating the Tangible and Intangible of Samoan Diaspora Social Space', *idea journal* 12, no. 1 (2012): 122–39, https://doi.org/10.37113/ideaj.v0i0.99.

50 'To draw using the relational line one must question what is to be recorded', Muller, 'Relational Drawing as Agency', 136; 138.

51 Paterson, Mulla and Macken, 'Drawing the Room', 271.

52 Marian Macken, Sarosh Mulla and Aaron Paterson, 'The Architectural Office Space as a Measuring Tool', in *The Routledge Companion to Architectural Drawings and Models: Translating, Collecting, Archiving and Displaying*, ed. Federica Goffi (Routledge, 2022), 308.

53 Yeoryia Manolopoulou and Níall McLaughlin, *Losing Myself: Bartlett Design Research Folios* (Bartlett School of Architecture, 2022), 64, https://www.bartlettdesignresearchfolios.com/manolopoulou-mclaughlin-losing-myself/.

54 Yeoryia Manolopoulou, 'Performing the Architectural Plan: Egocentric and Allocentric Drawing', in *Arts & Dementia: Interdisciplinary Perspectives*, ed. Ruth Mateus-Berr and L. Vanessa Gruber (De Gruyter, 2021), 304.

55 Manolopoulou, 'Performing the Architectural Plan', 300.

56 Manolopoulou, 'Performing the Architectural Plan', 300.

57 Manolopoulou, 'Performing the Architectural Plan', 300.

58 Manolopoulou, 'Performing the Architectural Plan', 300.

59 Thomas-Bernard Kenniff and Carole Lévesque, 'Introduction', in *In Drawing: Inquiry, Time, Dialogue and Materiality*, ed. Thomas-Bernard Kenniff and Carole Lévesque (Bureau d'étude de pratiques indisciplinées, 2024), 9.

60 'Dreamlife of Driverless Cars', ScanLAB Projects, https://scanlabprojects.co.uk/work/dreamlife-of-driverless-cars/, accessed 2 October 2024; see also Carl Douglas, 'Point/Cloud: Diffusive Spatial Imaginaries', *idea journal* 20, no. 1 (2023): 26–43, https://doi.org/10.37113/ij.v20i01.496.

61 Geoff Manaugh, 'The Dream Life of Driverless Cars', *New York Times* 11 November (2015), https://www.nytimes.com/2015/11/15/magazine/the-dream-life-of-driverless-cars.html.

62 Jussi Parikka, 'On Seeing Where There's Nothing to See: Practices of Light Beyond Photography', in *Photography Off the Scale: Technologies and Theories of the Mass Image*, ed. Tomás Dvořák and Jussi Parikka (Edinburgh University Press, 2021), 198.

63 Mark Rakatansky, 'The Gestic Body of Architecture', in *The Body*, ed. Andrew Benjamin (Academy Editions; Ernst & Sohn, 1993), 70.

64 Jasmine Benyamin, 'Analog Dreams', in *306090: A Journal of Emergent Architecture and Design* 11, ed. Emily Abruzzo, Eric Ellingsen and Jonathan D. Solomon (306090, School of Architecture, Princeton University, 2007), 90.

65 John Hejduk, 'Afterword', in *Stanley Tigerman: Buildings and Projects, 1966–1989*, ed. Stanley Tigerman and Sarah Mollman Underhill (Rizzoli, 1989), 259.

66 Marco Frascari, 'Horizons at the Drafting Table: Filarete and Steinberg', in *Chora* 5, ed. Alberto Pérez Gómez and Stephen Parcell (McGill-Queen's University Press; Combined Academic, 2007), 184.

67 Maxine Sheets-Johnstone, *The Primacy of Movement* (John Benjamins Publishing, 1999), 359.

68 David Turnbull, 'Performance and Narrative, Bodies and Movement in the Construction of Places and Objects, Spaces and Knowledge: The Case of Maltese Megaliths', *Theory, Culture and Society* 19, no. 5/6 (2002): 135.

69 Carlo Rovelli, *The Order of Time*, trans. Simon Carnell and Erica Segre (Riverhead, 2018), 97.

70 Lindsay Jones, *The Hermeneutics of Sacred Architecture, Volume 2: Hermeneutical Calisthenics: A Morphology of Ritual-Architectural Priorities* (Harvard University Press, 2000), 263.

71 Miwon Kwon, *One Place After Another: Site-Specific Art and Locational Identity* (MIT Press, 2002), 3.

72 Kwon, *One Place After Another*, 91.

73 Kwon, *One Place After Another*, 110.

Chapter 4

1 Thordis Arrhenius, 'Discourse', in *Exhibiting Architecture: Place and Displacement*, ed. Thordis Arrhenius, Mari Lending, Wallis Miller and Jérémie Michael McGowan (Lars Müller Publishers, 2014), 17.

2 Rolf Hughes, 'Exposition and the Staging of Encounter: On Assessing Unconventional Research Outputs', in *Creative Practice Inquiry in Architecture*, ed. Ashley Mason and Adam Sharr (Routledge, 2023), 15.

3 'To exhibit means to choose, to display, to present a sample or an example.' Klaus Franck (1961), cited in Mark Beck, 'The Exhibition and the Display', in *Exhibition*, ed. Lucy Steeds (MIT Press, 2014), 27. Fleur Watson describes the emergence of the 'new curator' who 'identifies and articulates creative content that explores progressive and experimental ideas' rather than simply organizing and choosing. Fleur Watson, *The New Curator: Exhibiting Architecture and Design* (Routledge, 2021), 13.

4 The 11th Venice Biennale of Architecture, curated by Aaron Betsky, recognized this challenge, focusing on 'architecture beyond building'. Contributions were termed site-specific installation, visions and experiments. International Architectural Exhibition, Biennale di Venezia, Aaron Betsky, *Out There: Architecture Beyond Building 1: Installations* (Rizzoli; Marsilio, 2008).

5 Claire Bishop, *Installation Art: A Critical History* (Tate Publishing, 2005), 6.

6 Simon Twose, Jules Moloney, Anastasia Globa and Lawrence Harvey, 'Drawing the Unfixed', *Interstices: Journal of Architecture and Related Arts* 21 (2022): 125, https://doi.org/10.24135/ijara.vi.694.

7 Bishop, *Installation Art*, 10.

8 Maurice Merleau-Ponty, *The Primacy of Perception, and Other Essays on Phenomenological Psychology, the Philosophy of Art, History, and Politics* (Northwestern University Press, 1964), 178; Bishop, *Installation Art*, 50.

9 'Although many artists and architects privilege phenomenological experience, they often offer the near-reverse: "experience" handed back to us as "atmosphere" or "affect" – that is, environments that confuse the actual with the virtual, or feelings that are hardly our own yet interpellate us nonetheless.' Hal Foster, 'The Art-Architecture Complex', in *Expanded Field: Architectural Installation Beyond Art: Berman + Burnham*, ed. Ila Berman and Douglas Burnham (Applied Research and Design Publishing, 2016), 357.

10 Miwon Kwon, *One Place After Another: Site-Specific Art and Locational Identity* (MIT Press, 2002), 3.

11 Kwon, *One Place After Another*, 3.

12 Mark Wigley, 'On Site', *Lotus International* 95 (1997): 119.

13 Kwon, *One Place After Another*, 2.

14 Sophia Banou, 'Installation/Drawing: Spaces of Drawing Between Art and Architecture', in *A Companion to Contemporary Drawing*, ed. Kelly Chorpening, Rebecca Fortnum and Dana Arnold (Wiley-Blackwell, 2020), 439.

15 Banou, 'Installation/Drawing', 432.

16 Nezar AlSayyad, *Cities and Caliphs: On the Genesis of Arab Muslim Urbanism* (Greenwood Press, 1991), 115–17.

17 Al-Tabari, *The History of al-Tabarī, 'Abbāsid Authority Affirmed* 28, ed. Ehsan Yar-Shater, trans. Jane Dammen McAuliffe (State University of New York Press, 1995), 245–6.

18 David Ross Scheer, *The Death of Drawing: Architecture in the Age of Simulation* (Routledge, 2014), 103.

19 Lifesize Plans, 'Lifesize Plans: About', https://lifesizeplans.com.au/about/, accessed 16 October 2024.

20 Banou, 'Installation/Drawing', 446. *On the Surface* was curated by Karen Kjaegaard and installed at the Aarhus School of Architecture.

21 Mark Dorrian, 'In the Abyss of the Infinitesimally Small', in *Micromegas > Powers of 10*, ed. Tracey Mackenna and Edwin Janssen (Dundee Contemporary Arts, 2013), 8.

22 Sue Gallagher, 'Home-Body: A Spatial Inquiry into Queer Material Imagination' (PhD diss., Auckland University of Technology, 2020), 107.

23 Gallagher, 'Home-Body', 108.

24 Bishop, *Installation Art*, 11.

25 James A. Craig and Matthew Ozga-Lawn, 'Models as Objects: The Installation as Architectural Encounter', *Architectural Design* 91, no. 3 (2021): 87.

26 Exhibited in *Time Space Existence: 4th edition*. 26 May–5 November 2018. https://timespaceexistence.com/past-editions/time-space-existence-2018.

27 Twose et al., 'Drawing the Unfixed', 129.

28 Simon Twose, Jules Moloney and Lawrence Harvey, 'Canyon: Experiments in Drawing a Landscape', *Drawing On: Journal of Architectural Research by Design* 3 (2019): 24, https://doi.org/10.24135/ijara.vi.694.

29 Twose et al., 'Drawing the Unfixed', 123; Simon Twose and Jules Moloney, 'Drawing Canyon: Sfumato Presences in Drawing and Landscape', *Interstices: Journal of Architecture and Related Arts* 19 (2019): 41, https://doi.org/10.26686/wgtn.12494156.

30 Twose et al., 'Drawing the Unfixed', 128.

31 Twose et al., 'Drawing the Unfixed', 126.

32 Francesca Hughes, *The Architecture of Error: Matter, Measure, and the Misadventures of Precision* (MIT Press, 2014), 41.

33 Twose et al., 'Drawing the Unfixed', 126.

34 Twose and Moloney, 'Drawing Canyon', 50.

35 Twose and Moloney, 'Drawing Canyon', 50.

36 David Brashears, *Last Climb: The Legendary Everest Expeditions of George Mallory* (National Geographic Society, 1999), 175.

37 In 2024, a team of climbers on Mount Everest also discovered, protruding from the ice in the melting glacier, a leather boot; inside was a frozen foot and sock, labelled with Irvine's name. Alexa Robles-Gil, 'Famous Explorer's Remains Discovered on Mount Everest Offer Clues in a Century-Long Mystery', *Smithsonian Magazine* 15 October (2024), https://www.smithsonianmag.com/smart-news/famous-explorers-remains-discovered-on-mount-everest-offer-clues-in-a-century-long-mystery-180985267/.

38 James A. Craig and Matthew Ozga-Lawn, 'Mallory's Ascent: Engaging the Space of Death Through Architectural Drawing', in *Expanding Fields of Architectural Discourse and Practice: Curated Works from the P.E.A.R. Journal*, ed. Matthew Butcher and Megan O'Shea (UCL Press, 2020), 347. See also Craig and Ozga-Lawn, 'Models as Objects', 87.

39 *The Epic of Everest* (1924; restored 2020) [Film], dir. John Baptist Lucius Noel.

40 James Craig and Matthew Ozga-Lawn, 'Everest Death Zone: Bodily Transgressions in Architectural Drawing', in *Mountains and Megastructures: Neo-Geologic Landscapes of Human Endeavour*, ed. Martin Beattie, Christos Kakalis and Matthew Ozga-Lawn (Springer, 2020), 126–7.

41 Craig and Ozga-Lawn, 'Mallory's Ascent', 350.

42 'Video: AirXY', M-A-D, Adigard+Salter, http://airxy.org/video.html, accessed 16 October 2024.

43 International Architectural Exhibition, Biennale di Venezia, *Out There*, 117.

44 International Architectural Exhibition, Biennale di Venezia, *Out There*, 117.

45 Aaron Paterson, Sarosh Mulla and Marian Macken, *Drawing Room*, Toi Moroki Centre of Contemporary Art, Christchurch, New Zealand (2020–1). The exhibition comprised of *Drawing Machine* (music and sound design by Fabio Morreale), *Edge of Shadow* (digital VR environment; sound design by Clovis McEvoy; research assistance by Xiaoling Cheng) and *Penumbral* (short film; directed by Brendan Donovan, Mulla and Paterson).

46 See, for instance, Robin Evans, *Translations from Drawing to Building and Other Essays* (AA Publications, 1997), 163–5.

47 Aaron Paterson, Sarosh Mulla and Marian Macken, *Drawing Room,* exhibition catalogue (Toi Moroki Centre of Contemporary Art Christchurch, 2021), 37–8.

48 Lynda Simmons, 'Interior Darkness/Contained Shadow', *Interstices: Journal of Architecture and Related Arts* 12 (2011): 127, https://doi.org/10.24135/ijara.v0i0.421.

49 Exhibited at Objectspace, Auckland (2018), https://www.objectspace.org.nz/exhibitions/penumbral-reflections/; a short film based on the work was included in *Drawing Room* (2020–1). See Crinkle Cut House and Waimataruru Houses: Pac Studio, https://www.pacstudio.nz/projects, accessed 30 November 2024.

50 Fred Sandback, 'Untitled', in *Fred Sandback* (Kunstmuseum Liechtenstein; Fruitmarket Gallery; Neue Galerie am Landesmuseum Joanneum; CAPC musée d'art contemporain; Hatje Cantz, 2005), 152, cited in Fred Sandback, *Fred Sandback: Vertical Constructions*, ed. David Gray and Lucas Zwirner (David Zwirner Books, 2017), 43.

51 Juhani Pallasmaa, *Eyes of the Skin* (Academy Editions, 1996), 44–5.

52 Maurice Merleau-Ponty, *Phenomenology of Perception*, trans. Donald A. Landes (Routledge, 2012), 140.

53 Elizabeth Grosz, *Volatile Bodies: Toward a Corporeal Feminism* (Indiana University Press, 1994), xi.

Chapter 5

1 Stephen Temple, *Making Thinking: Beginning Architectural Design Education* (Kendall Hunt Publishing, 2011), 3.

2 'Studio was articulated as both spatial and temporal; "the studio" is a place, but "studio" is also a time and a set of practices including those of thinking, making and talking.' Kathy Waghorn and Ross T. Smith, 'Bodies+ Spaces', *idea journal* 9, no. 1 (2009): 163.

3 Edward Robbins, *Drawing and Architectural Practice Revisited* (MIT Press, 1994), 300.

4 Donald Schön, *The Design Studio: An Exploration of Its Traditions and Potentials* (RIBA Publications, 1985), 26.

5 James A. Craig, Christos P. Kakalis and Matthew Ozga-Lawn, 'On Disjointed Bodies: Emergent Spaces Between the Body and Screen in Pandemic-Era Architectural Education', *Charrette* 7, no. 1 (2021): 50.

6 This does not mean there is no value in online learning tools: '[T]he internet can become a studio for a radically anarchic and pataphysical practice of education, but only if we struggle against the dominance of the metaphysics of learning that continues to colonize the broader expanse of the postdigital world, threatening to reduce educational life to mere economic management.' Peter B. Hyland and Tyson E. Lewis, *Studious Drift: Movements and Protocols for a Postdigital Education* (University of Minnesota Press, 2022), 2.

7 Craig, Kakalis and Ozga-Lawn, 'On Disjointed Bodies', 43.

8 For an expanded overview on this, see Lee Stickells, 'Little Living Labs: 1970s Student Design-Build and the Objects of Experimental Lifestyles', in *Architectural Education Through Materiality: Pedagogies of 20th Century Design*, ed. Elke Couchez and Rajesh Heynickx (Routledge, 2022), 126–45. Stickells notes that none of these projects was repeated.

9 Beatriz Colomina, 'Learning from Global Tools', in *Global Tools 1973–1975: When Education Coincides with Life*, ed. Valeria Borgonuovo and Silvia Franceschini (Nero Editions, 2018); cited in Stickells, 'Little Living Labs', 128.

10 Bill McKay, 'The Counter-Culture and Its Containment: The Loose Years', in *The Auckland School: 100 Hundred Years of Architecture and Planning*, ed. Julia Gatley and Lucy Treep (Auckland School of Architecture and Planning, 2017), 85.

11 McKay, 'The Counter-Culture and Its Containment', 76.

12 Silvia Franceschini, 'The Body as an Ultimate Form of Architecture: Global Tools Body Workshops', in *Architectural Education Through Materiality: Pedagogies of*

20th Century Design, ed. Elke Couchez and Rajesh Heynickx (Routledge, 2022), 149–63.

13 Franceschini, 'Body as an Ultimate Form of Architecture', 153–4.

14 Franceschini, 'Body as an Ultimate Form of Architecture', 161.

15 Raumlaborberlin, 'Floating University Berlin 2018 – An Illustrated Report', https://raumlabor.net/floating-university-berlin-book/, accessed 31 October 2024.

16 An example of an ongoing academic practice of live projects can be seen in the Timber Research Structures series led by Professor Andrew Barrie at Waipapa Taumata Rau University of Auckland since 2012. Andrew Barrie Lab, 'Timber Research Structures: Summerhill Series', https://www.andrewbarrielab.com/timber-research-structures-summerhill, accessed 30 November 2024.

17 Stickells, 'Little Living Labs', 142.

18 Course delivered at Xi'an Jiaotong-Liverpool University (XJTLU), Suzhou, China, 2013–15; course coordinator Marian Macken; cohort sizes between 168 and 196.

19 Marian Macken, 'Wear: trio/triform/triplet >> Wearing a Conversation' (course briefing, XJTLU, 2016).

20 Karen A. Franke and R. Bianca Lepori, *Architecture from the Inside Out* (Wiley-Academy, 2007), 18–19.

21 Course delivered at Waipapa Taumata Rau University of Auckland, 2017–24; course coordinator Marian Macken; cohort sizes between 125 and 160 students; second of three architectural media core courses.

22 Paul Emmons and Carolina Dayer, 'Toward Performative Architectural Drawing: Paul Klee's Enacted Lines', in *Architecture as Performing Art*, ed. Marcia Feuerstein and Gray Read (Ashgate, 2013), 46.

23 These three possibilities of research in art and design (about, for and through) are set out in Christopher Frayling, 'Research in Art and Design', *Royal College of Art Research Papers 1*, no. 1 (1993/4): 1–5.

24 Sarah Barnes, 'Architectural Drawing: It Is an Action' (MArch(Prof) thesis, University of Auckland, 2022), 14; supervised by Marian Macken.

25 Barnes, 'Architectural Drawing', 34.

26 Maurice Merleau-Ponty, *Phenomenology of Perception* (Routledge, 2012), 95.

27 Barnes, 'Architectural Drawing', 77.

28 These drawings in particular recall and adapt William Anastasi's Subway Drawing (Way to John Cage) (1988).

29 J. Ferreira Crespo, 'Walking: Drawing the City with the Body', in *Drawing (...) City (...) Body, Dwelling on Earth: Imagined-Architectures: Architectural Graphic Representation and Other Images*, ed. Pedro António Janeiro (CRC Press, 2019), 95; cited in Barnes, 'Architectural Drawing', 169–70.

30 Barnes, 'Architectural Drawing', 98.

31 Examiner's report, 20 April 2023, Barnes, 'Architectural Drawing'.

32 Barnes, 'Architectural Drawing', 218.

33 Nicola Chang, 'The Learning Anatomy: A Performative School of Architectural Bodies' (MArch(Prof) thesis, University of Auckland, 2022); supervised by Marian Macken.

34 Nikolaus Gansterer, Emma Cocker and Mariella Greil, eds, *Choreo-Graphic Figures: Deviations from the Line* (De Gruyter, 2017); Lucy McRae, 'Work', https://www.lucymcrae.net/home, accessed 15 July 2025.

35 Chang, 'Learning Anatomy', 3.

36 Chang, 'Learning Anatomy', 19.

37 Chang, 'Learning Anatomy', 116.

38 Chang, 'Learning Anatomy', 136.

39 Chang, 'Learning Anatomy', 136.

40 Matthew Davis, 'Object-Sketch: An Exploration into the Materiality of Digital Fabrication and 3D Printed Ceramics' (MDes thesis., Auckland University of Technology, 2024), 13; supervised by Carl Douglas and James Charlton.

41 Davis, 'Object-Sketch', 58.

42 Davis, 'Object-Sketch', 30.

43 Davis, 'Object-Sketch', 17.

44 The simulation used flocking algorithms in Grasshopper, a visual coding tool built into Rhino 3D.

45 Davis, 'Object-Sketch', 62.

46 Stan Allen, *Practice: Architecture, Technique, Representation* (Routledge, 2009), xiii.

47 Elizabeth B. Hatz, 'Building, Seeing, Delaying', in *Drawing in Architecture and Education Research*, ed. Heike Biechteler, Dieter Dietz, Johannes Käferstein and Jonathan Sergison (Park Books, 2023), 55.

48 Isak Boardman, Abstract, 'Dwelling in Drawing: A House for Father and Son' (MArch diss., KTH, 2020), https://kth.diva-portal.org/smash/record.jsf?pid=diva2%3A1442846&dswid=-1022, accessed 18 November 2024,.

49 Boardman, 'Dwelling in Drawing'.

50 Eva Flores and Ricardo Prats, *Thought by Hand: The Architecture of Flores & Prats* (Arquine, 2014), 292.

51 This method derives from the PhD by Practice programme at RMIT University (where Prats completed her doctorate). Eva Prats, 'To Observe with the Client, to Draw with the Existing: Three Cases of Architecture Dealing with the As-Found' (PhD thesis, RMIT, 2019). Mike Davis, Esther Mecredy and Alessandro Melis, 'Material Openings: Ark and the Materiality of the Vessel', *Drawing On: Journal of Architectural Research by Design* 2 (2017): 61–76.

52 Gretchen Wilkins, 'The Agency of Making', *AD* 93, no. 3 (2023): 122.

53 Temple, *Making Thinking*, 4–5.

54 Jane Rendell, 'Seeing Time/Writing Place', in *The Pits*, ed. Janet Hodgson (SWPA Publishing, 2006), 185.

55 Hyland and Lewis, *Studious Drift*, 19.

56 Kathy Waghorn and Nick Sargent, 'The City as a School: An Urban Pedagogy', *Drawing On: Journal of Architectural Research by Design* 8 (2020): 123.

Conclusion

1 Among the key texts from this period, we have particularly drawn from Stan Allen, *Practice: Architecture, Technique and Representation* (Routledge, 2009); Robin Evans, *The Projective Cast* (MIT Press, 1995); Alberto Pérez-Gómez and Louise Pelletier, *Architectural Representation and the Perspective Hinge* (MIT Press, 1997); and Catherine Ingraham, *Architecture and the Burdens of Linearity* (Yale University Press, 1998).

2 Paul Emmons, *Drawing Imagining Building: Embodiment in Architectural Design Practices* (Routledge, 2019), 12. By 'thickness' we also invoke Geertz's 'thick descriptions' and Eklund's 'thick concepts'. Glifford Geertz, *The Interpretation of Culture: Selected Essays* (Basic Books, 1973), 3; Matti Eklund, 'What Are Thick Concepts?' *Canadian Journal of Philosophy* 41, no. 1 (2011): 25–49, https://doi.org/10.1353/cjp.2011.0007.

3 Michael Young, *Reality Modeled After Images: Architecture and Aesthetics After the Digital Image* (Routledge, 2021); María Fullaondo, *The Colors of Rhetoric: Places of Invention in the Spatial Realm* (Applied Research and Design Publishing, 2022).

4 Federica Goffi, ed., *The Routledge Companion to Architectural Drawings and Models: Translating, Collecting, Archiving and Displaying* (Routledge, 2022); Hamza Shaik, *Drawing Attention: Architecture in the Age of Social Media* (RIBA Publishing, 2023).

5 Mark Dorrian, Riet Eeckhout and Arnaud Hendrickx, *Drawing Architecture: Conversations on Contemporary Practice* (Lund Humphries, 2022); Martin Søberg and Anna Hougaard, eds, *The Artful Plan: Architectural Drawing Reconfigured* (Birkhäuser, 2020).

6 Drawing Matter, 'Drawing Matter: About', https://drawingmatter.org/about/, accessed 3 December 2024.

7 Tim Ingold, *Lines: A Brief History* (Routledge, 2007), 75.

8 Ingold, *Lines*, 75; Paul Klee, *Notebooks. Volume 1. The Thinking Eye*, ed. Jürg Spiller, trans. Ralph Manheim (Lund Humphries, 1961), 105.

9 David Rosand, *Drawing Acts: Studies in Graphic Expression and Representation* (Cambridge University Press, 2002), 1.

10 Elizabeth Diller, recounting her experience as an architecture student at Cooper Union, New York, writes that in this teaching, drawings were constructed analogously to buildings: 'For us, drawing and building were parallel manifestations of architecture, in no particular order. Each had distinct attributes that were missing from the other. While drawings lacked the sensation of full-scale immersion of their three-dimensional counterpart, they revealed organizations and forces hidden in actual space.' Elizabeth Diller, 'Autobiographical Notes', in *The Activist Drawing: Retracing Situationist Architectures from Constant's New Babylon to Beyond*, ed. Catherine de Zegher and Mark Wigley (MIT Press, 2001), 131.

11 Evans, *The Projective Cast*, 366.

12 Emmons, *Drawing Imagining Building*, 217; David Ross Scheer, *The Death of Drawing: Architecture in the Age of Simulation* (Routledge, 2014), 2.

13 Ingraham, *Architecture and the Burdens of Linearity*.

14 It recalls paintings by Anmatyerr artist Emily Kam Kngwarray, such as *Seeds of Abundance* (1990) and *Ntang* (1990). Kngwarray's art stems from *awely*, 'a composite embodied, embedded, and multimodal aesthetic of place-based, patrilineal songs, associated ochre breast, upper arms and body marks, ritual object forms and dance cycles'. Jennifer Loureide Biddle, 'Emily Kam Kngwarray', *Visual Anthropology Review* (2024), https://doi.org/10.1111/var.70004.

15 Raukura Turei and Jade Townsend, 'Whenua as a Catalyst for Reconnection', *The Art Paper* 18 November (2024), https://www.the-art-paper.com/journal/raukura-turei-jade-townsend-season. Turei considers her painting practice as being tactile and present in a way architectural practice can rarely be.

16 James Scott, *Seeing like a State: How Certain Schemes to Improve the Human Condition Have Failed* (Yale University Press, 1999), 80.

17 Scott, *Seeing like a State*, 36.

18 Giselle Byrnes writes that the erection of survey poles 'signified an explicit act of possession. They were symbols of authority, statements declaring the presence of a system or individual in whom authority, however temporary, was invested.' Giselle Byrnes, *Boundary Markers: Land Surveying and the Colonisation of New Zealand* (Bridget Williams Books, 2001), 180.

19 Anna Lowenhaupt Tsing, 'On Non-Scalability: The Living World Is Not Amenable to Precision-Nested Scales', *Common Knowledge* 18, no. 3 (2012): 523, https://doi.org/10.1215/0961754X-1630424.

20 Elizabeth Grosz, 'Notes Towards a Corporeal Feminism', *Australian Feminist Studies* 2, no. 5 (1997): 8, https://doi.org/10.1080/08164649.1987.9961562.

21 Jean-Luc Nancy, *The Pleasure in Drawing* (Fordham University Press, 2013), 1.

BIBLIOGRAPHY

Abraham, Raimund. *Ungebaut/Unbuilt*. Forum AR/GE Kunst, 1986.

Al-Tabari. *The History of al-Tabarī, 'Abbāsid Authority Affirmed* 28. Edited by Ehsan Yar-Shater. Translated by Jane Dammen McAuliffe. State University of New York Press, 1995.

Alberti, Leon Battista. *Delineation of the City of Rome.* Edited by Mario Carpo and Francesco Furlan. Casa Editrice Leo C. Olschki, 2005.

Alberti, Leon Battista. *Descriptio Urbis Romae*. Edited by Jean-Yves Boriaud and Francesco Furlan. Casa Editrice Leo C. Olschki, 2005.

Alberti, Leon Battista. *On Painting and On Sculpture: The Latin Texts of De Pictura and De Statua.* Edited by Cecil Grayson. Phaidon, 1972.

Alberti, Leon Battista. *On the Art of Building in Ten Books.* Translated by Joseph Rykwert, Neil Leach and Robert Tavernor. MIT Press, 1988.

Alexander, Zeynep Çelik. *Kinaesthetic Knowing: Aesthetics, Epistemology, Modern Design.* University of Chicago Press, 2017.

Allen, Jim. *The Skin of Years: Interviews with Jim Allen by Phil Dadson and Tony Green.* Clouds and Michael Lett, 2014.

Allen, Laura and Luke Pearson, eds. *Drawing Futures: Speculations in Contemporary Drawing for Art and Architecture*. UCL Press, University College London, 2016.

Allen, Matthew. *Flowcharting: From Abstractionism to Algorithmics in Art and Architecture*. Gta Verlag, 2023.

Allen, Stan. *Practice: Architecture, Technique, Representation.* Routledge, 2009.

AlSayyad, Nezar. *Cities and Caliphs: On the Genesis of Arab Muslim Urbanism.* Greenwood Press, 1991.

Anderson, Kristi. *The Geometry of an Art: The History of the Mathematical Theory of Perspective from Alberti to Monge.* Springer, 2007.

Andrew Barrie Lab. 'Timber Research Structures: Summerhill Series'. https://www.andrewbarrielab.com/timber-research-structures-summerhill, Accessed 30 November 2024.

Arrhenius, Thordis, Mari Lending, Wallis Miller and Jérémie Michael McGowan, eds. *Exhibiting Architecture: Place and Displacement*. Lars Müller, 2014.

Badiou, Alain. 'Drawing'. *Lacanian Ink* 28 (2006): 42–9.

Baek, Jin. 'From the "Topos of Nothingness" to the "Space of Transparency": Kitarō Nishida's Notion of Shintai and Its Influence on Art and Architecture'. *Philosophy East and West* 58, no. 1 (2008): 83–107.

Banou, Sophia. 'nstallation/Drawing: Spaces of Drawing Between Art and Architecture'. In *A Companion to Contemporary Drawing*, edited by Kelly Chorpening, Rebecca Fortnum and Dana Arnold. Wiley-Blackwell, 2020.

Barnes, Sarah. 'Architectural Drawing: It is an Action'. MArch(Prof) thesis, University of Auckland, 2022.

Benyamin, Jasmine. 'Analog Dreams'. In *306090: A Journal of Emergent Architecture and Design* 11, edited by Emily Abruzzo, Eric Ellingsen and Jonathan D. Solomon. 306090, School of Architecture, Princeton University, 2007.

Berman, Ila and Douglas Burnham. *Expanded Field: Architectural Installation Beyond Art: Berman + Burnham*. Applied Research and Design Publishing, 2016.

Biddle, Jennifer Loureide. 'Emily Kam Kngwarray'. *Visual Anthropology Review* (2024). https://doi.org/10.1111/var.70004.

Bishop, Claire. *Installation Art: A Critical History*. Tate Publishing, 2005.

Blau, Eve and Edward Kaufman, eds. *Architecture and its Image: Four Centuries of Architectural Representation: Works from the Collection of the Canadian Centre for Architecture.* Canadian Centre for Architecture; MIT Press, 1989.

Boardman, Isak. 'Dwelling in Drawing: A House for Father and Son'. MArch thesis, KTH, 2020. https://urn.kb.se/resolve?urn=urn:nbn:se:kth:diva-276818, Accessed 18 November 2024.

Bochner, Mel, Barry Le Va, Dorothea Rockburne, Richard Tuttle: Exhibition, 8 January–16 February 1975. Contemporary Arts Center, 1975.

Borges, Jorge Luis. *Collected Fictions*. Penguin Books, 1999.

Boys, Jos, ed. *Disability, Space, Architecture: A Reader*. Routledge, 2017.

Brown, Elspeth H. *The Corporate Eye: Photography and the Rationalization of American Commercial Culture, 1884-1929*. Johns Hopkins University Press, 2005.

Bruno, Giuliana. *Atlas of Emotion: Journeys in Art, Architecture, and Film*. Verso, 2018.

Butler, Cornelia H. and M. Catherine de Zegher. *On Line: Drawing Through the Twentieth Century*. Museum of Modern Art, 2011.

Byrnes, Giselle. *Boundary Markers: Land Surveying and the Colonisation of New Zealand*. Bridget Williams Books, 2015.

Carpo, Mario. *The Digital Turn in Architecture 1992–2012*. Wiley, 2013.

Carpo, Mario. *The Second Digital Turn: Design Beyond Intelligence.* MIT Press, 2017.

Carroll, Lewis. *Sylvie and Bruno Concluded*. Macmillan, 1894.

Casey, Edward S. *The Fate of Place: A Philosophical History*. University of California Press, 1997.

Cathcart, James, Frank Fantauzzi and Terence Van Elslander. *Gravity: Pamphlet Architecture 25*. Princeton Architectural Press, 2003.

Chang, Nicola. 'The Learning Anatomy: A Performative School of Architectural Bodies'. MArch(Prof) thesis, University of Auckland, 2022.

Cixous, Hélène. 'Without End No State of Drawingness No, Rather: The Executioner's Taking Off'. Translated by Catherine A.F. MacGillivray. *New Literary History* 24, no. 1 (1993): 87–103.

Clark, Justine. 'Smudges, Smears and Adventitious Marks'. *Interstices: Journal of Architecture and Related Arts* 4 (1999): 1–8. https://interstices.ac.nz/index.php/Interstices/article/view/568.

Coen, Deborah J. *Climate in Motion: Science, Empire and the Problem of Scale*. University of Chicago, 2018.

Contreras, Javier Fernandez. *The Miralles Projection: Thinking and Representation in the Architecture of Enric Miralles*. ORO Editions, 2020.

Cooper, M. A. R. 'Robert Hooke's Work as Surveyor for the City of London in the Aftermath of the Great Fire. Part One: Robert Hooke's First Surveys for the City of London'. *Notes and Records of the Royal Society of London* 51, no. 2 (1997): 161–74.

Cosgrove, Ben. 'Behind the Picture: Picasso Draws with Light'. *Life*. https://www.life.com/arts-entertainment/behind-the-picture-picasso-draws-with-light/, Accessed 1 October 2024.

Craig, James A. and Matthew Ozga-Lawn. 'Everest Death Zone: Bodily Transgressions in Architectural Drawing'. In *Mountains and Megastructures: Neo-Geologic Landscapes of Human Endeavour*, edited by Martin Beattie, Christos Kakalis and Matthew Ozga-Lawn. Springer, 2020.

Craig, James A. and Matthew Ozga-Lawn. 'Mallory's Ascent: Engaging the Space of Death through Architectural Drawing'. In *Expanding Fields of Architectural Discourse and Practice: Curated Works from the P.E.A.R. Journal*, edited by Matthew Butcher and Megan O'Shea. UCL Press, 2020.

Craig, James A. and Matthew Ozga-Lawn. 'Models as Objects: The Installation as Architectural Encounter'. *Architectural Design* 91, no. 3 (2021): 82–7. https://doi.org/10.1002/ad.2696.

Craig, James A., Christos P. Kakalis and Matthew Ozga-Lawn. 'On Disjointed Bodies: Emergent Spaces between the Body and Screen in Pandemic-era Architectural Education'. *Charrette* 7, no. 1 (2021): 41–58.

Crespo, J. Ferreira. 'Walking: Drawing the City with the Body'. In *Drawing (...) City (...) Body, Dwelling on Earth: Imagined-Architectures: Architectural Graphic Representation and Other Images*, edited by Pedro António Janeiro. CRC Press, 2019.

Damisch, Hubert. *The Origin of Perspective*. Translated by John Goodman. MIT Press, 1994.

Davis, Matthew. 'Object-Sketch: An Exploration into the Materiality of Digital Fabrication and 3D Printed Ceramics'. MDes thesis, Auckland University of Technology, 2024.

Davis, Mike, Esther Mecredy and Alessandro Melis. 'Material Openings: Ark and the Materiality of the Vessel'. *Drawing On. Journal of Architectural Research by Design* 2 (2017): 61–76.

Derrida, Jacques. *Memoirs of the Blind: The Self-Portrait and Other Ruins*. Translated by Pascale-Anne Brault and Michael Nas. University of Chicago Press, 1993.

Descartes, René. *The Philosophical Writings of Descartes, Vol. II*. Translated by John Cottingham, Robert Stoothoff and Dugald Murdoch. Cambridge University Press, 1984.

Diller, Elizabeth. 'Autobiographical Notes'. In *The Activist Drawing: Retracing Situationist Architectures from Constant's New Babylon to Beyond*, edited by Catherine De Zegher and Mark Wigley. MIT Press, 2001.

Diller, Elizabeth and Ricardo Scofidio. *Flesh: Architectural Probes*. Princeton Architectural Press, 1994.

Dorrian, Mark. 'Adventure on the Vertical'. *Cabinet Magazine*, no. 44 (2012). https://www.cabinetmagazine.org/issues/44/dorrian.php.

Dorrian, Mark. 'In the Abyss of the Infinitesimally Small'. In *Micromegas > Powers of 10*, edited by Tracey Mackenna and Edwin Janssen. Dundee Contemporary Arts, 2013.

Dorrian, Mark and Frédéric Poussin. *Seeing from Above: The Aerial View in Visual Culture*. I.B. Tauris, 2013.

Dorrian, Mark, Riet Eeckhout and Arnaud Hendrickx, eds. *Drawing Architecture: Conversations on Contemporary Practice*. Lund Humphries, 2022.

Douglas, Carl. 'Do Not Scale: A Lament for Design Drawing'. *TRACEY* 16, no. 1 (2022): 1–13.

Douglas, Carl. 'Point/Cloud: Diffusive Spatial Imaginaries'. *idea journal* 20, no. 1 (2023): 26–43. https://doi.org/10.37113/ij.v20i01.496.

Downton, Peter. *Design Research. Revised Kindle Edition*. Elizabeth James Productions, 2013.

Drawing Matter. 'Drawing Matter: About'. https://drawingmatter.org/about/, Accessed 3 December 2024.

Dürer, Albrecht. *The Painter's Manual: A Manual of Measurement of Lines, Areas, and Solids by Means of Compass and Ruler Assembled by Albrecht Dürer for the Use of All Lovers of Art with Appropriate Illustrations Arranged to be Printed in the Year MDXXV*. Translated by Walter L. Strauss. Abaris Books, 1977.

Dürer, Albrecht. *Vier Bücher von Menschlicher Proportion.* Edited by Berthold Hinz. Akademie Verlag, 2011.

Dvořák, Tomás and Jussi Parikka, eds. *Photography Off the Scale: Technologies and Theories of the Mass Image.* Edinburgh University Press, 2021.

Eklund, Matti. 'What Are Thick Concepts?' *Canadian Journal of Philosophy* 41, no. 1 (2011): 25–49. https://doi.org/10.1353/cjp.2011.0007.

Emmons, Paul. *Drawing Imagining Building: Embodiment in Architectural Design Practices*. Routledge, 2019.

Emmons, Paul. 'Drawn to Scale: The Imaginative Inhabitation of Architectural Drawings'. In *From Models to Drawings: Imagination and Representation*, edited by Marco Frascari, Jonathan Hale and Bradley Starkey. Routledge, 2007.

Emmons, Paul. 'Size Matters: Virtual Scale and Bodily Imagination in Architectural Drawing'. *ARQ* 9, no. 3–4 (2005): 227–35. https://doi.org/10.1017/S135913550500028X.

Emmons, Paul and Carolina Dayer. 'Toward Performative Architectural Drawing: Paul Klee's Enacted Lines'. In *Architecture as Performing Art*, edited by Marcia Feuerstein and Gray Read. Ashgate, 2013.

Engels-Schwarzpaul, Anna-Christina and Michael A. Peters, eds. *Of Other Thoughts: Non-traditional Ways to the Doctorate*. Sense, 2013.

Euclid. *Euclid's Elements of Geometry*. Edited by Richard Fitzpatrick. Richard Fitzpatrick, 2008.

Evans, Robin. *The Projective Cast: Architecture and Its Three Geometries*. MIT Press, 1995.

Evans, Robin. *Translations from Drawing to Building and Other Essays*. AA Publications, 1997.

Flatt, Russ. 'Te ahua, te wa, te atea'. https://www.russflatt.com/te-ahua-te-wa-te-atea, Accessed 22 September 2024.

Flores, Eva and Ricardo Prats. *Thought by Hand: The Architecture of Flores & Prats*. Arquine, 2014.

Foote, Nancy. 'Drawing the Line'. *Artforum* 14, no. 9 (1976): 54–6.

Foster, Hal, ed. *Vision and Visuality.* Bay Press, 1988.

Frampton, Kenneth and Sylvia Kolbowski, eds. *Idea as Model*. Rizzoli International Publications, 1981.

Francesca, Piero della. *De Prospectiva Pingendi*. Edited by Chiara Gizzi. Fondazione Università Ca' Foscari, 2016. http://doi.org/10.14277/978-88-6969-099-0.

Franceschini, Silvia. 'The Body as an Ultimate Form of Architecture: Global Tools Body Workshops'. In *Architectural Education Through Materiality: Pedagogies of 20th Century Design*, edited by Elke Couchez and Rajesh Heynickx. Routledge, 2022.

Franck, Karen A. and R. Bianca Lepori. *Architecture from the Inside Out*. Wiley-Academy, 2007.

Frascari, Marco. *Eleven Exercises in the Art of Architectural Drawing: Slow Food for the Architect's Imagination.* Routledge, 2011.

Frascari, Marco. 'Horizons at the Drafting Table: Filarete and Steinberg'. In *Chora* 5, edited by Alberto Péréz-Gómez and Stephen Parcell. McGill-Queen's University Press; Combined Academic, 2007.

Frascari, Marco. 'Lines as Architectural Thinking'. *Architectural Theory Review* 14, no. 3 (2009): 200–12. https://doi.org/10.1080/13264820903341605.

Frayling, Christopher. 'Research in Art and Design'. *Royal College of Art Research Papers* 1, no. 1 (1993/4): 1–5.

Fullaondo, María. *The Colors of Rhetoric: Places of Invention in the Spatial Realm*. Applied Research and Design Publishing, 2022.

Galfetti, Gustau Gili, ed. *Allan Wexler*. GG Portfolio, 1998.

Gallagher, Sue. 'Home-Body: A Spatial Inquiry into Queer Material Imagination'. PhD thesis, Auckland University of Technology, 2020. https://hdl.handle.net/10292/13755.

Galloway, Alexander. *Uncomputable: Play and Politics in the Long Digital Age*. Verso Books, 2021.

Gansterer, Nikolaus, Emma Cocker and Mariella Greil, eds. *Choreo-Graphic Figures: Deviations from the Line*. De Gruyter, 2017.

Ganz, Kate. *Drawing Then: Innovation and Influence in American Drawings of the Sixties*. Dominique Lévy, 2016.

Geertz, Glifford. *The Interpretation of Culture: Selected Essays.* Basic Books, 1973.

Gil, José. 'Paradoxical Body'. Translated by André Lepecki. *The Drama Review* 50, no. 4 (2006): 21–35. https://doi.org/10.1162/dram.2006.50.4.21.

Gilbreth, Frank Bunker and Robert Thurston Kent. *Motion Study: A Method for Increasing the Efficiency of the Workman*. D. Van Nostrand Company, 1911.

Gissen, David. *The Architecture of Disability: Buildings, Cities, and Landscapes Beyond Access*. University of Minnesota Press, 2023.

Goffi, Federica, ed. *The Routledge Companion to Architectural Drawings and Models: Translating, Collecting, Archiving and Displaying*. Routledge, 2022.

Goodman, Nelson. *Languages of Art: An Approach to a Theory of Symbols*. Bobbs-Merrill, 1968.

Graham, Laurel D. 'Domesticating Efficiency: Lillian Gilbreth's Scientific Management of Homemakers, 1924–1930'. *Signs* 24, no. 3 (1999): 633–75. https://doi.org/10.1086/495368.

Grant, Elizabeth, Kelly Greenop, Albert L. Refiti and Daniel J. Glenn, eds. *The Handbook of Contemporary Indigenous Architecture*. Springer, 2018.

Gregory, Nuala. 'Encountering Drawing'. *Educational Philosophy and Theory* 49, no. 13 (2017): 1236–45. https://doi.org/10.1080/00131857.2017.1336919.

Gross, Béatrice, ed. *Sol LeWitt*. Centre Pompidou-Metz; M-Museum; Terra Foundation for American Art, 2012.

Grosz, Elizabeth. 'Notes Towards a Corporeal Feminism'. *Australian Feminist Studies* 2, no. 5 (1997): 1–16. https://doi.org/10.1080/08164649.1987.9961562.

Grosz, Elizabeth. *Space, Time and Perversion: Essays on the Politics of Bodies*. Routledge, 1995.

Grosz, Elizabeth. *Volatile Bodies: Toward a Corporeal Feminism*. Indiana University Press, 1994.

Haraway, Donna J. *Staying with the Trouble: Making Kin in the Chthulucene*. Duke University Press, 2016.

Hatz, Elizabeth B. 'Building, Seeing, Delaying'. *In Drawing in Architecture and Education Research*, edited by Heike Biechteler, Dieter Dietz, Johannes Käferstein and Jonathan Sergison. Park Books, 2023.

Healy, Claire and Sean Cordeiro. 'The Cordial Home Project'. https://claireandsean.com/large-scale-projects/the-cordial-home-project, Accessed 29 November 2024.

Heathcote, Edwin. *Imre Makovecz: The Wings of the Soul*. Academy Editions, 1997.

Herbert, Daniel M. *Architectural Study Drawings*. Van Nostrand Reinhold, 1993.

Hill, Jonathan. 'Weathering the Barcelona Pavilion'. *Journal of Architecture* 7, no. 4 (2002): 319–27. https://doi.org/10.1080/13602360320000040820.

Hoffman, Dan. *Architecture Studio: Cranbrook Academy of Art, 1986–1993*. Rizzoli, 1994.

Hoffman, Dan. 'Unbuilding and the Recovery of Craft in Architecture: Cranbrook Department of Architecture 1986–1996'. *Architectural Design* 93, no. 3 (2023): 70–7. http://dx.doi.org/10.1002/ad.2936.

Hooke, Robert. *Micrographia: Or Some Physiological Descriptions of Minute Bodies Made by Magnifying Glasses with Observations and Inquiries Thereupon*. Jo. Martyn and Ja. Allestry, 1665.

Hubert, Christian. 'The Ruins of Representation Revisited'. *OASE* 84 (2011): 11–16.

Hughes, Francesca. *The Architecture of Error: Matter, Measure, and the Misadventures of Precision*. MIT Press, 2014.

Hughes, Rolf. 'Exposition and the Staging of Encounter: On Assessing Unconventional Research Outputs'. In *Creative Practice Inquiry in Architecture*, edited by Ashley Mason and Adam Sharr. Routledge, 2023.

Husserl, Edward. *The Crisis of European Sciences and Transcendental Phenomenology: An Introduction to Phenomenological Philosophy*. Translated by David Carr. Northwestern University, 1970.

Hyland, Peter B. and Tyson E. Lewis. *Studious Drift: Movements and Protocols for a Postdigital Education*. University of Minnesota Press, 2022.

Ingold, Tim. *Lines: A Brief History*. Routledge, 2007.

Ingraham, Catherine. *Architecture and the Burdens of Linearity*. Yale University Press, 1998.

Ingraham, Catherine. 'Lines and Linearity: Problems in Architectural Theory'. In *Drawing/Building/Text: Essays in Architectural Theory*, edited by Andrea Kahn. Princeton Architectural Press, 1991.

International Architectural Exhibition, Biennale di Venezia, Aaron Betsky. *Out There: Architecture Beyond Building 1: Installations*. Rizzoli; Marsilio, 2008.

Jardine, Lisa. *The Curious Life of Robert Hooke: The Man Who Measured London*. Harper Perennial, 2003.

Jenner, Ross. 'Thought Out of Bounds: Theory and Practice in Architecture Doctorates'. In *Of Other Thoughts: Non-Traditional Ways to the Doctorate: A Guidebook for Candidates and Supervisors*, edited by A-Chr. Engels-Schwarzpaul and Michael A. Peters. Sense Publishers, 2013.

Jones, Amelia. *Body Art: Performing the Subject*. University of Minnesota Press, 1998.

Jones, Lindsay. *The Hermeneutics of Sacred Architecture, Volume 2: Hermeneutical Calisthenics: A Morphology of Ritual-Architectural Priorities*. Harvard University Press, 2000.

Kahn, Andrea and Carol J. Burns, eds. *Site Matters: Strategies for Uncertainty Through Planning and Design*. Routledge, 2021.

Kauffman, Jordan. *Drawing on Architecture: The Object of Lines, 1970–1990*. MIT Press, 2018.

Kauffman, Jordan. 'Model Values'. In *Valuing Architecture: Heritage and the Economics of Culture*, edited by Ashley Paine, Susan Holden and John Macarthur. Valiz, 2020.

Kaye, Nick. *Site-Specific Art: Performance, Place and Documentation*. Routledge, 2000.

Kayser, Christine Vial and Sylvie Coëllier, eds. *Installation Art as Experience of Self, in Space and Time*. Vernon Press, 2021.

Kenniff, Thomas-Bernard and Carole Lévesque, eds. *In Drawing: Inquiry, Time, Dialogue and Materiality*. Bureau d'étude de pratiques indisciplinées, 2024.

Kiesler, Frederick. 'Pseudo-Functionalism in Modern Architecture'. *Partisan Review* 16, no. 7 (1949): 733–42.

Kilday, Bill. *Never Lost Again: The Google Mapping Revolution That Sparked New Industries and Augmented Our Reality*. HarperCollins, 2018.

Klee, Paul. *Notebooks. Volume 1. The Thinking Eye*, edited by Jürg Spiller. Translated by Ralph Manheim. Lund Humphries, 1961.

Krauss, Rosalind. *Line as Language: Six Artists Draw*. Princeton University Press, 1974.

Krauss, Rosalind. *The Originality of the Avant-Garde and Other Modernist Myths*. MIT Press, 1985.

Kulper, Perry. 'The Labor of Architectural Drawing'. *Figure* 1 (2014): 10–11.

Kwon, Miwon. 'One Place after Another: Notes on Site Specificity'. *October* 80 (Spring 1997): 85–110. https://doi.org/10.2307/778809.

Kwon, Miwon. *One Place after Another: Site-Specific Art and Locational Identity*. MIT Press, 2002.

Lambert, Léopold and Minh-Ha T. Pham. 'Spinoza in a T-Shirt'. *The New Inquiry* 1 July (2015). https://thenewinquiry.com/spinoza-in-a-t-shirt/.

Leatherbarrow, David. 'Showing What Otherwise Hides Itself: On Architectural Representation'. *Harvard Design Magazine* Fall (1998): 55.

Lee, Pamela M. Object *to Be Destroyed: The Work of Gordon Matta-Clark*. MIT Press, 2000.

Lefaivre, Liane and Alexander Tzonis. 'Die Maschine Im Architektonischen Denken: The Machine in Architectural Thinking'. *Daidalos* 18 (1985): 16–26.

Leonard, Robert. 'Jim Allen: Contact'. *Art New Zealand* 95 (2000). https://robertleonard.org/jim-allen-contact/.

Lifesize Plans. 'Lifesize Plans: About'. https://lifesizeplans.com.au/about/, Accessed 16 October 2024.

Linzey, Mike. 'Architectural Drawings Do Not Represent'. *Interstices: A Journal of Architecture and Related Arts* 11 (2010): 31–9. https://doi.org/10.24135/ijara.v0i0.378.

Lissitzky, El. 'The Film of El's Life'. In *El Lissitzky: Life, Letters, Text*, edited by Sophie Lissitzky-Küppers. Thames and Hudson, 1968.

Lodder, Christina. 'Seeing Red: Lissitzky's Abstract Cabinet and the Ideology of Display'. *Umění Art* 6 (1999): 512–20.

Lovatt, Anna. 'Dorothea Rockburne: Intersection'. *October* 122 (2007): 47. https://doi.org/10.1162/octo.2007.122.1.31.

Lovatt, Anna. 'Ideas in Transmission: LeWitt's Wall Drawings and the Question of the Medium'. *Tate Papers* 14 (1 October 2010). https://www.tate.org.uk/research/tate-papers/14/ideas-in-transmission-lewitt-wall-drawings-and-the-question-of-medium.

Macken, Marian. 'Documenting Interiority / Inhabiting Duration'. In *The Interior Architecture Theory Reader*, edited by Gregory Marinic. Routledge, 2018.
Macken, Marian, Sarosh Mulla and Aaron Paterson. 'The Architectural Office Space as a Measuring Tool'. In *The Routledge Companion to Architectural Drawings and Models: Translating, Collecting, Archiving and Displaying*, edited by Federica Goffi. Routledge, 2022.
Macy, Christine. 'The Authority of Architectural Drawing'. In *Knowledge and/or/of Experience: The Theory of Space in Art and Architecture*, edited by John Macarthur. Institute of Modern Arts, 1993.
M-A-D, Adigard+Salter. 'Video: AirXY'. http://airxy.org/video.html, Accessed 16 October 2024.
Maier, Jessica. *Rome Measured and Imagined: Early Modern Maps of the Eternal City*. University of Chicago Press, 2015.
Maillet, Arnaud. *The Claude Glass: Use and Meaning of the Black Mirror in Western Art*. Zone Books, 2009.
Manaugh, Geoff. 'The Dream Life of Driverless Cars'. *New York Times* 11 November (2015). https://www.nytimes.com/2015/11/15/magazine/the-dream-life-of-driverless-cars.html.
Manolopoulou, Yeoryia. *Architectures of Chance*. Routledge, 2013.
Manolopoulou, Yeoryia. 'Performing the Architectural Plan: Egocentric and Allocentric Drawing'. In *Arts & Dementia: Interdisciplinary Perspectives*, edited by Ruth Mateus-Berr and L. Vanessa Gruber. De Gruyter, 2021.
Manolopoulou, Yeoryia. 'Unformed Drawing; Notes, Sketches, and Drawings'. *The Journal of Architecture* 10, no. 5 (2005): 517–25. https://doi.org/10.1080/13602360500462323.
Manolopoulou, Yeoryia and Níall McLaughlin. *Losing Myself: Bartlett Design Research Folios*. Bartlett School of Architecture, 2022. https://www.bartlettdesignresearchfolios.com/manolopoulou-mclaughlin-losing-myself/.
May, John. *Signal. Image. Architecture*. Columbia University Press, 2019.
McEwen, Indra Kagis. *All the King's Horses: Vitruvius in an Age of Princes*. MIT Press, 2023.
McEwen, Indra Kagis. *Vitruvius: Writing the Body of Architecture*. MIT Press, 2002.
McKay, Bill. 'The Counter-Culture and its Containment: The Loose Years'. In *The Auckland School: 100 Years of Architecture and Planning*, edited by Julia Gatley and Lucy Treep. Auckland School of Architecture and Planning, 2017.
McRae, Lucy. 'Work'. https://www.lucymcrae.net/home. Accessed 15 July 2025.
MeGo. 'Floor Plan, Melissa Gould'. http://www.megophone.com/floorplan.html, Accessed 16 September 2024.
Meister, Anna-Maria. 'Ernst Neufert's 'Lebensgestaltungslehre': Formatting Life beyond the Built'. *BJHS Themes* 5 (2020): 167–85, https://doi.org/10.1017/bjt.13.
Meredith, Michael and Hilary Sample, eds. *An Unfinished Encyclopedia of Scale Figures without Architecture*. MIT Press, 2019.
Merleau-Ponty, Maurice. *Phenomenology of Perception*. Translated by Donald A. Landes. Routledge, 2012.
Merleau-Ponty, Maurice. *The Primacy of Perception, and Other Essays on Phenomenological Psychology, the Philosophy of Art, History, and Politics*. Northwestern University Press, 1964.
Milne, Victoria. 'Bringing the Past to Light'. *Metropolis Magazine* January/February (1994): 33.

Müller, Karamia. ‘Relational Drawing as Agency: Negotiating the Tangible and Intangible of Samoan Diaspora Social Space’. *idea journal* 12, no. 1 (2012): 122–39, https://doi.org/10.37113/ideaj.v0i0.99.

Nancy, Jean-Luc. *The Pleasure in Drawing*. Fordham University Press, 2013.

Nelson, Maggie. *Like Love: Essays and Conversations*. Penguin Random House, 2024.

Neufert, Ernst and Peter Neufert. *Architects’ Data. Third Edition*. Edited by Bousmaha Baiche and Nicholoas Walliman. Blackwell Science, 2008.

Obrist, Hans Ulrich. ‘Installations are the Answer, What is the Question?’ *Oxford Art Journal,* 24, no. 2 (2001): 93–101. https://doi.org/10.1093/oxartj/24.2.93.

Oxford English Dictionary. https://www.oed.com/dictionary, Accessed 3 December 2024.

Pai, Hyungmin. *The Portfolio and the Diagram: Architecture, Discourse, and Modernity in America*. MIT Press, 2002.

Pallasmaa, Juhani. *Eyes of the Skin*. Academy Editions, 1996.

Paterson, Aaron, Sarosh Mulla and Marian Macken. *Drawing Room*. Toi Moroki Centre of Contemporary Art Christchurch, 2021.

Paterson, Aaron, Sarosh Mulla and Marian Macken. ‘Drawing the Room | Drawing within the Room’. *Drawing: Research, Theory, Practice* 5, no. 2 (2020): 261–75. https://doi.org/10.1386/drtp_00036_1.

Péréz-Gómez, Alberto. ‘The Revelation of Order: Perspective and Architectural Representation’. In *This is Not Architecture: Media Constructions*, edited by Kester Rattenbury. Routledge, 2002.

Péréz-Gomez, Alberto and Louise Pelletier. *Architectural Representation and the Perspective Hinge*. MIT Press, 1997.

Petherbridge, Deanna. *The Primacy of Drawing: Histories and Theories of Practice.* Yale University Press, 2010.

Piccarolo, Gaia. ‘Textile Architecture’. *Lotus International* 176 (2024): 48–53.

Prats, Eva. ‘To Observe with the Client, to Draw with the Existing: Three Cases of Architecture Dealing with the As-Found’. PhD thesis, RMIT University, 2019.

Rakatansky, Mark. ‘The Gestic Body of Architecture’. In *The Body*, edited by Andrew Benjamin. Academy Editions; Ernst & Sohn, 1993.

Raumlaborberlin. ‘Floating University Berlin 2018 – An Illustrated Report’. https://raumlabor.net/floating-university-berlin-book/, Accessed 31 October 2024.

Reiss, Julie H. *From Margin to Center: The Spaces of Installation Art*. MIT Press, 2000.

Rendell, Jane. ‘Seeing Time/ Writing Place’. In *The Pits*, edited by Janet Hodgson. SWPA Publishing, 2006.

Riahi, Pari, Laure Katsaros and Michael T. Davis. *Exactitude: On Precision and Play in Contemporary Architecture*. University of Massachusetts, 2022.

Riley, Terrence. *Light Construction*. Museum of Modern Art, 1995.

Robbins, Edward. *Drawing and Architectural Practice Revisited*. MIT Press, 1994.

Robles-Gil, Alexa. ‘Famous Explorer’s Remains Discovered on Mount Everest Offer Clues in a Century-Long Mystery’. *Smithsonian Magazine* 15 October (2024). https://www.smithsonianmag.com/smart-news/famous-explorers-remains-discovered-on-mount-everest-offer-clues-in-a-century-long-mystery-180985267/.

Rockburne, Dorothea. ‘Notes to Myself on Drawing’. *Flash Art* April (1974): 66.

Rosand, David. *Drawing Acts: Studies in Graphic Expression and Representation.* Cambridge University Press, 2002.

Rovelli, Carlo. *The Order of Time*. Translated by Simon Carnell and Erica Segre. Riverhead, 2018.

Ryan, Zoë, ed., *As Seen: Exhibitions that Made Architecture and Design History*. Art Institute of Chicago, 2017.

Said, Edward. *Orientalism: Western Conceptions of the Orient*. Penguin Books, 1995.

Sandback, Fred. *Fred Sandback: Räume Zeichnen / Drawing Spaces; Arbeiten auf Papier / Works on Paper*. Kerber, 2011.

Sandback, Fred. *Fred Sandback: Vertical Constructions,* edited by David Gray and Lucas Zwirner. David Zwirner Books, 2017.

ScanLAB Projects. 'Dreamlife of Driverless Cars'. https://scanlabprojects.co.uk/work/dreamlife-of-driverless-cars/, Accessed 2 October 2024.

Scheer, David Ross. *The Death of Drawing: Architecture in the Age of Simulation*. Routledge, 2014.

Schmidlin, Laurence. 'Drawing as a Tool for Mapping the Body in Space'. In *Body, Space, and Place in Collective and Collaborative Drawing; Drawing Conversations II*, edited by Jill Journeaux, Helen Gørrill and Sara Reed. Cambridge Scholars Publishing, 2020.

Schön, Donald. *The Design Studio: An Exploration of its Traditions and Potentials*. RIBA Publications, 1985.

Schön, Donald. *The Reflective Practitioner: How Professionals Think in Action*. Routledge, 1994.

Scott, James. *Seeing Like a State: How Certain Schemes to Improve the Human Condition Have Failed*. Yale University Press, 1999.

Seidenberg, Abraham. 'The Ritual Origin of Geometry'. *Archive for History of Exact Sciences* 1, no. 5 (1962): 488–527.

Shaik, Hamza. *Drawing Attention: Architecture in the Age of Social Media*. RIBA Publishing, 2023.

Sheets-Johnstone, Maxine. *The Primacy of Movement*. John Benjamins Publishers, 1999.

Simmons, Lynda. 'Interior Darkness/Contained Shadow'. *Interstices:* Journal *of Architecture and Related Arts* 12 (2011): 126–30. https://doi.org/10.24135/ijara.v0i0.421.

Simone, Ashley, ed. *Michael Webb: Two Journeys*. Lars Müller Publishers, 2018.

Søberg, Martin and Anna Hougaard, eds. *The Artful Plan: Architectural Drawing Reconfigured*. Birkhäuser, 2020.

Sorkin, Michael. 'Michael Webb, Temple Island: AA Exhibition Gallery, 1–30 May 1987'. *AA Files* 15 (1987): 67–72.

Spector, Naomi. 'Dorothea Rockburne'. In *Mel Bochner, Barry Le Va, Dorothea Rockburne, Richard Tuttle: Exhibition, Jan 8–Feb 16, 1975*. Contemporary Arts Center, 1975.

Spiller, Neil. *Visionary Architecture: Blueprints of the Modern Imagination*. Thames and Hudson, 2006.

Steeds, Lucy, ed. *Exhibition*. MIT Press, 2014.

Stickells, Lee. 'Little Living Labs: 1970s Student Design-build and the Objects of Experimental Lifestyles'. In *Architectural Education Through Materiality: Pedagogies of 20th Century Design*, edited by Elke Couchez and Rajesh Heynickx. Routledge, 2022.

Stout, Katharine. *Contemporary Drawing from the 1960s to Now*. Tate Publ., 2014.

Sutherland, Ivan Edward. 'Sketchpad, A Man-Machine Graphical Communication System'. PhD thesis, Massachusetts Institute of Technology, 1963.

Temple, Stephen. *Making Thinking: Beginning Architectural Design Education*. Kendall Hunt Publishing, 2011.

The Epic of Everest (1924; 2020), [Film]. Directed by John Baptist Lucius Noel.

The Stage of Drawing: Gesture and Act, Selected from the Tate Collection by Avis Newman, Curated by Catherine De Zegher. Tate Publishing; The Drawing Center, 2003.

Tigerman, Stanley and Sarah Mollman Underhill. *Stanley Tigerman: Buildings and Projects, 1966–1989*. Rizzoli, 1989.

Till, Jeremy. *Architecture Depends*. MIT Press, 2009.

Tomkins, Calvin. 'The Maverick'. *New Yorker* 7 July (1997): 38–45.

Tsing, Anna Lowenhaupt. 'On Non-Scalability: The Living World Is Not Amenable to Precision-Nested Scales'. *Common Knowledge* 18, no. 3 (2012): 505–24. https://doi.org/10.1215/0961754X-1630424.

Turei, Raukura and Jade Townsend. 'Whenua as a Catalyst for Reconnection'. *The Art Paper* 18 November (2024). https://www.the-art-paper.com/journal/raukura-turei-jade-townsend-season.

Turnbull, David. 'Performance and Narrative, Bodies and Movement in the Construction of Places and Objects, Spaces and Knowledge: The Case of Maltese Megaliths'. *Theory, Culture and Society* 19, no. 5–6 (2002): 125–43. https://doi.org/10.1177/02632760276189.

Twose, Simon and Jules Moloney. 'Drawing Canyon: Sfumato Presences in Drawing and Landscape'. *Interstices: Journal of Architecture and Related Arts* 19 (2019): 41–53. https://doi.org/10.24135/ijara.v0i0.554.

Twose, Simon, Jules Moloney and Lawrence Harvey. 'Canyon: Experiments in Drawing a Landscape'. *Drawing On: Journal of Architectural Research by Design* 3 (2019): 123–36. https://doi.org/10.24135/ijara.vi.694.

Twose, Simon, Jules Moloney, Anastasia Globa and Lawrence Harvey. 'Drawing the Unfixed'. *Interstices: Journal of Architecture and Related Arts* 21 (2022): 123–36. https://doi.org/10.24135/ijara.vi.694.

Vasari, Giorgio. *Lives of the Artists, Volume 1*. Penguin, 1987.

Vervoort, Stefaan. 'Scale Models and Postmodernism: Revisiting Idea as Model (1976-1981)'. *Architectural Theory Review,* 24, no. 3 (2020): 224–40.

Vesely, Dalibor. *Architecture in the Age of Divided Representation: The Question of Creativity in the Shadow of Production*. MIT Press, 2004.

Vidler, Anthony. 'Diagrams of Diagrams: Architectural Abstraction and Modern Representation'. *Representations* 72 (2000): 1–20. https://doi.org/10.2307/2902906.

Vidler, Anthony. *The Architectural Uncanny: Essays in the Modern Unhomely*. MIT Press, 1992.

Waghorn, Kathy and Nick Sargent. 'The City as a School: An Urban Pedagogy'. *Drawing On. Journal of Architectural Research by Design* 8 (2020): 121–34.

Waghorn, Kathy and Ross T. Smith. 'Bodies+ Spaces'. *idea journal* 9, no. 1 (2009): 162–71. https://doi.org/10.37113/ideaj.vi0.151.

Walker, Matthew F. 'The Limits of Collaboration: Robert Hooke, Christopher Wren and the Designing of the Monument to the Great Fire of London'. *Notes and Records of the Royal Society of London* 65, no. 2 (2011): 121–43.

Wall, Cynthia Sundberg. *The Prose of Things: Transformations of Description in the Eighteenth Century*. University of Chicago Press, 2014.

Webb, Michael, Michael Sorkin and Lebbeus Woods. *Temple Island: A Study*. Architectural Association, 1987.

Wigley, Mark. *Cutting Matta-Clark: The Anarchitecture Investigation*. Lars Müller Publishers, 2018.
Wigley, Mark. 'On Site'. *Lotus International* 95 (1997): 118–31.
Wilkins, Gretchen. 'The Agency of Making'. *AD* 93, no. 3 (2023): 121–7. https://doi.org/10.1002/ad.2942.
Williams, Kim and Michael J. Ostwald, eds. *Architecture and Mathematics from Antiquity to the Future: Vol. 2: The 1500s to the Future*. Springer, 2015.
Williams, Kim, Lionel March and Stephen R. Wassell, eds. *The Mathematical Works of Leon Battista Alberti*. Birkhäuser, 2010.
Williams, Todd and Billie Tsien. 'Slowness'. *G*, no. 9 (1999): 132–6.
Wilson, Andrew. 'Sol LeWitt Interviewed'. *Art Monthly*, no. 164 (March 1993): 3–9.
Wood, Peter. 'Drawing the Line: A Working Epistemology for the Study of Architectural Drawing'. PhD thesis, University of Auckland, 2002.
Yaneva, Albena. 'Scaling Up and Down: Extraction Trials in Architectural Design'. *Social Studies of Science* 35 (2005): 867–94.
Yaneva, Albena. *The Making of a Building: A Pragmatist Approach to Architecture.* Peter Lang, 2009.
Young, Michael. *Reality Modeled after Images: Architecture and Aesthetics after the Digital Image*. Routledge, 2021.
YouTube. 'Ivan Sutherland Sketchpad Demo 1963'. https://youtu.be/6orsmFndx_o, Accessed 24 September 2024.
YouTube. 'Rachael Lincoln performs Trisha Brown's 'Man Walking Down the Side of a Building''. https://www.youtube.com/watch?v=TbpBHTPODB4, Accessed 11 November 2024.
Zack, Maria. 'Robert Hooke's Fire Monument: Architecture as a Scientific Instrument'. In *Architecture and Mathematics from Antiquity to the Future: Vol. 2: The 1500s to the Future*, edited by Kim Williams and Michael J. Ostwald. Springer, 2015.

INDEX

Note: Page numbers in *italics* refer to figures and "n" refers to note numbers.